Ordinary Differential Equations

in each case (C and C' are real numbers):

(a) $\dot{x} = x - t$, R^2, $1 + t + Ce^t$, R;

(b) $\dot{x} = x^2$, R^2, $(C - t)^{-1}$, $(-\infty, C)$
0, R
$(C' - t)^{-1}$, (C', ∞);

(c) $\dot{x} = -x/t$, $\{(t, x) | t \neq 0\}$, C/t, $(-\infty, 0)$
C'/t, $(0, \infty)$;

(d) $\dot{x} = 2x^{1/2}$, $\{(t, x) | x \geqslant 0\}$, $\begin{cases} 0, & (-\infty, C) \\ (t - C)^2, & (C, \infty) \end{cases}$
0, R;

(e) $\dot{x} = 2xt$, R^2, Ce^{t^2}, R;

(f) $\dot{x} = -x/\tanh t$, $\{(t, x) | t \neq 0\}$, $C/\sinh t$, $(-\infty, 0)$
$C'/\sinh t$, $(0, \infty)$.

The existence of solutions is determined by the properties of X. The following proposition is stated without proof (Petrovski, 1966).

Proposition 1.1.1. If X is continuous in an open domain, $D' \subseteq D$, then given any pair $(t_0, x_0) \in D'$, there exists a solution $x(t)$, $t \in I$, of $\dot{x} = X(t, x)$ such that $t_0 \in I$ and $x(t_0) = x_0$.

For example, consider

$$\dot{x} = 2|x|^{1/2}, \tag{1.2}$$

where $D = \mathrm{R}^2$. Any pair (t_0, x_0) with $x_0 \geqslant 0$ is given by $(t_0, x(t_0))$ when $x(t)$ is the solution

$$x(t) = \begin{cases} 0, & t \in (-\infty, C) \\ (t - C)^2, & t \in (C, \infty) \end{cases} \tag{1.3}$$

and $C = t_0 - \sqrt{x_0}$. A solution can similarly be found for pairs (t_0, x_0) when $x_0 < 0$.

Observe that Proposition 1.1.1 does *not* exclude the possibility that $x(t_0) = x_0$ for more than one solution $x(t)$. For example, for (1.2) infinitely many solutions $x(t)$ satisfy $x(t_0) = 0$; namely every solution of the form (1.3) for which $C > t_0$ and the solution $x(t) \equiv 0$.

The following proposition gives a sufficient condition for each pair in D' to occur in one and only one solution of (1.1).

Proposition 1.1.2. If X and $\partial X / \partial x$ are continuous in an open domain $D' \subseteq D$, then given any $(t_0, x_0) \in D'$ there exists a *unique* solution $x(t)$ of $\dot{x} = X(t, x)$ such that $x(t_0) = x_0$.

Notice that, while $X = 2|x|^{1/2}$ is continuous on $D\,(= \mathrm{R}^2)$ $\partial X/\partial x$ $(= |x|^{-1/2}$ for $x > 0$ and $-|x|^{-1/2}$ for $x < 0)$ is continuous only on $D' = \{(t, x)|x \neq 0\}$; it is undefined for $x = 0$. We have already observed that the pair $(t_0, 0)$, $t_0 \in \mathrm{R}$ occurs in infinitely many solutions of $\dot{x} = 2|x|^{1/2}$.

On the other hand, $X(t, x) = x - t$ and $\partial X/\partial x = 1$ are continuous throughout the domain $D = \mathrm{R}^2$. Any (t_0, x_0) occurs in one and only one solution of $\dot{x} = x - t$; namely

$$x(t) = 1 + t + Ce^t \tag{1.4}$$

when $C = (x_0 - t_0 - 1)e^{-t_0}$.

Weaker sufficient conditions for existence and uniqueness do exist (Petrovski, 1966). However, Propositions 1.1.1 and 1.1.2 illustrate the kind of properties required for $X(t, x)$.

1.1.2 *Geometrical representation*

A solution $x(t)$ of $\dot{x} = X(t, x)$ is represented geometrically by the graph of $x(t)$. This graph defines a *solution curve* in the t, x-plane.

If X is continuous in D, then Proposition 1.1.1 implies that the solution curves fill the region D of the t, x-plane. This follows because each point in D must lie on at least one solution curve. The solutions of the differential equation are, therefore, represented by a *family of solution curves* in D (see Figs. 1.1–1.8).

If both X and $\partial X/\partial x$ are continuous in D then Proposition 1.1.2 implies that there is a *unique* solution curve passing through every point of D (see Figs. 1.1–1.6).

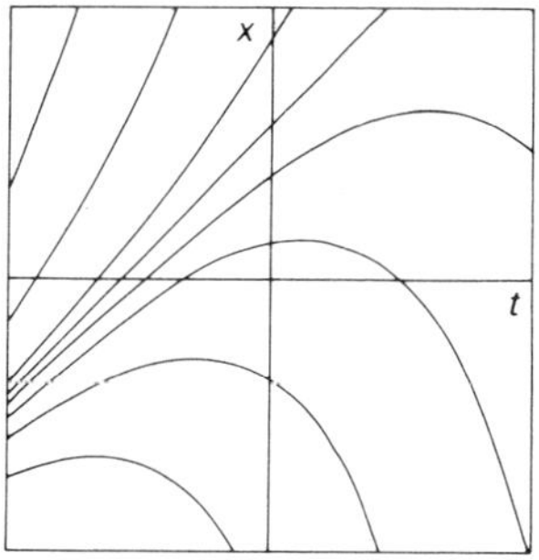

Fig. 1.1. $\dot{x} = x - t$.

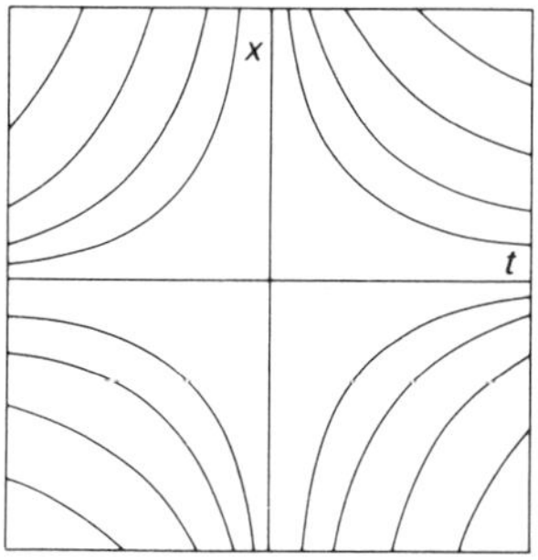

Fig. 1.2 $\dot{x} = -x/t, t \neq 0$.

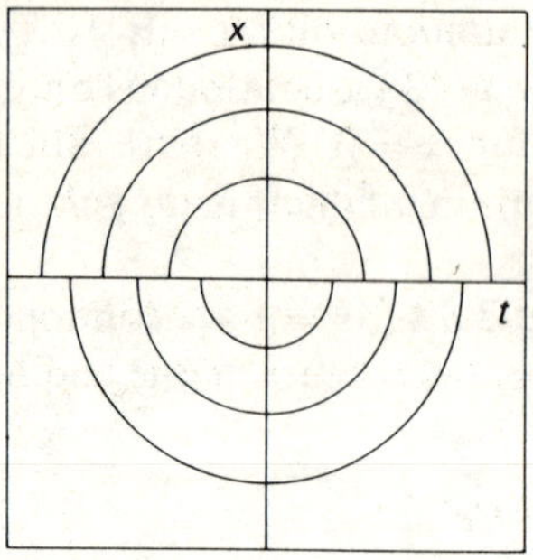

Fig. 1.3. $\dot{x} = -t/x$.

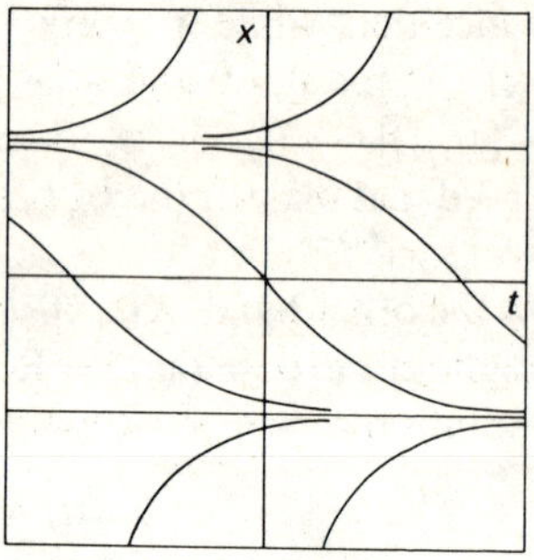

Fig. 1.4. $\dot{x} = \frac{1}{2}(x^2 - 1)$.

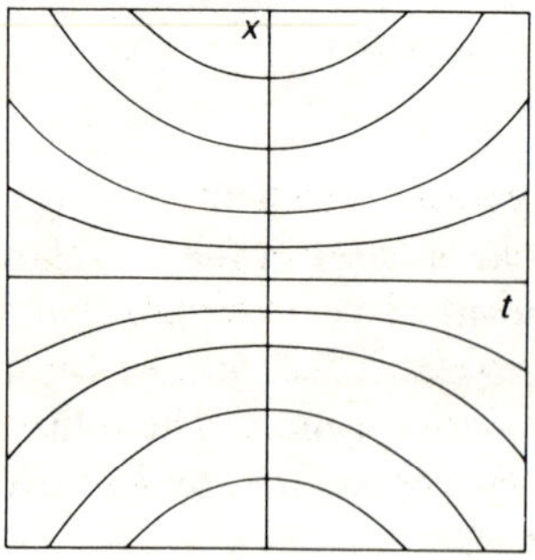

Fig. 1.5. $\dot{x} = 2xt$.

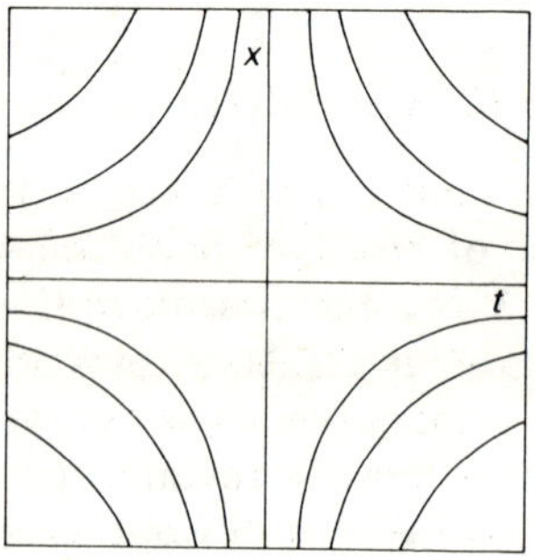

Fig. 1.6. $\dot{x} = -x/\tanh t, t \neq 0$.

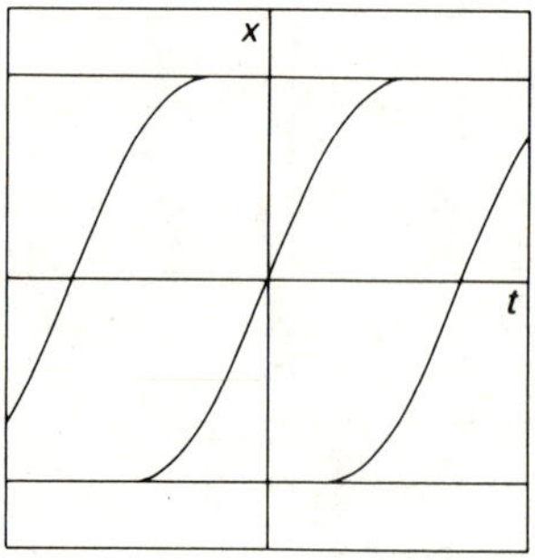

Fig. 1.7. $\dot{x} = \sqrt{(1-x^2)}, |x| \leqslant 1$,

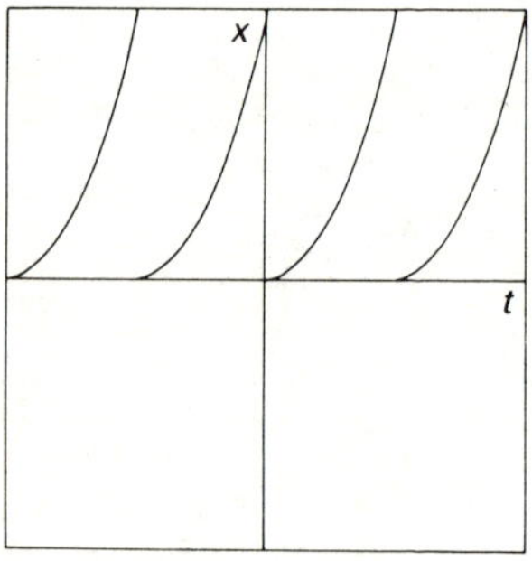

Fig. 1.8. $\dot{x} = 2x^{1/2}, x \geqslant 0$.

Observe that the families of solution curves in Figs. 1.2 and 1.6 bear a marked resemblance to one another. Every solution curve in one figure has a counterpart in the other; they are similar in shape, have the same asymptotes, etc., but they are *not* identical curves. The relationship between these two families of solution curves is an example of what we call *qualitative equivalence* (see Sections 1.3, 2.4 and 3.3). We say that the *qualitative behaviour* of the solution curves in Fig. 1.2 is the same as those in Fig. 1.6.

Accurate plots of the solution curves are not always necessary to obtain their qualitative behaviour; a *sketch* is often sufficient. We can sometimes obtain a sketch of the family of solutions curves directly from the differential equation.

Example 1.1.1. Sketch the solution curves of the differential equation

$$\dot{x} = t + t/x \tag{1.5}$$

in the region D of the t, x-plane where $x \neq 0$.

Solution. We make the following observations.

(a) The differential equation gives the slope of the solution curves at all points of the region D. Thus, in particular, the solution curves cross the curve $t + t/x = k$, a constant, with slope k. This curve is called the *isocline* of slope k. The set of isoclines, obtained by taking different real values for k, is a family of hyperbolae

$$x = \frac{t}{k - t}, \tag{1.6}$$

with asymptotes $x = -1$ and $t = k$. A selection of these isoclines is shown in Fig. 1.9.

(b) The sign of $\ddot{x}$ determines where in D the solution curves are concave and convex. If $\ddot{x} > 0$ (< 0) then $\dot{x}$ is increasing (decreasing) with t and the solution curve is said to be *convex* (*concave*). The region D can therefore be divided into subsets on which the solution curves are either concave or convex separated by boundaries where $\ddot{x} = 0$. For (1.5) we find

$$\ddot{x} = x^{-3}(x + 1)(x - t)(x + t) \tag{1.7}$$

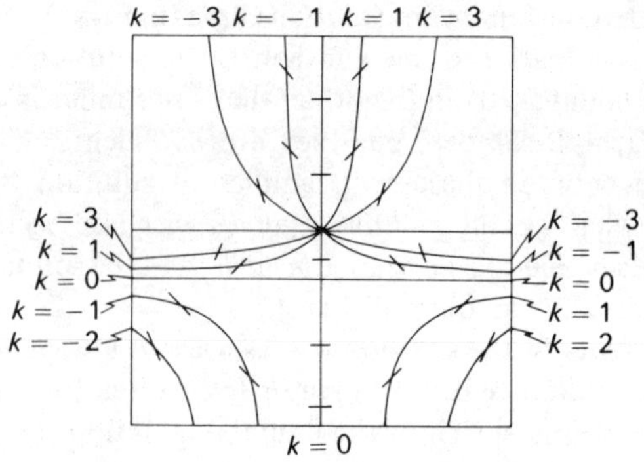

Fig. 1.9. Selected isoclines for the equation $\dot{x} = t + t/x$. The short line segments on the isoclines have slope k and indicate how the solution curves cross them.

and D can be split up into regions $P(\ddot{x} > 0)$ and $N(\ddot{x} < 0)$ as shown in Fig. 1.10.

(c) The isoclines are symmetrically placed relative to $t = 0$ and so there must also be symmetry of the solution curves. The function $X(t, x) = t + t/x$, satisfies $X(-t, x) = -X(t, x)$ and thus if $x(t)$ is a solution to $\dot{x} = X(t, x)$ then so is $x(-t)$ (see Exercise 1.5).

These three observations allow us to produce a sketch of the solution curves for $\dot{x} = t + t/x$ (see Fig. 1.11). Notice that both $X(t, x) = t + t/x$ and $\partial X/\partial x = -t/x^2$ are continuous on $D = \{(t, x) | x \neq 0\}$, so there is a unique solution curve passing through each point of D. □

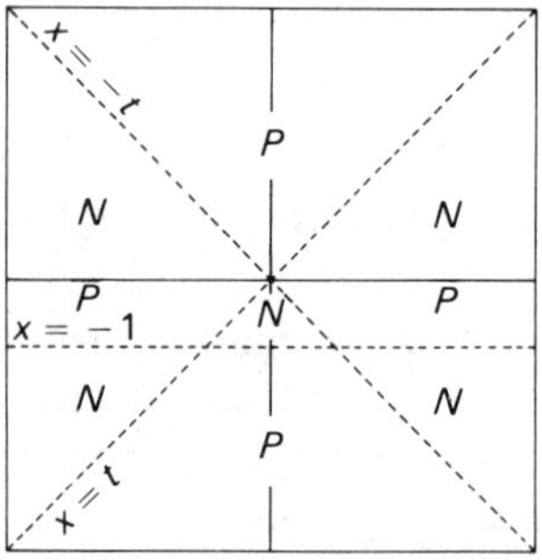

Fig. 1.10. Regions of convexity (P) and concavity (N) for solutions of $\dot{x} = t + t/x$.

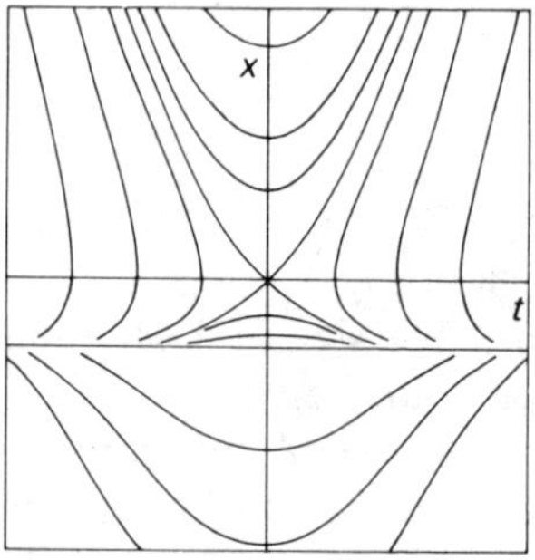

Fig. 1.11. The solution curves of the differential equation $\dot{x} = t + t/x$ in the t, x-plane.

It is possible to find the solutions of

$$\dot{x} = t + t/x \tag{1.8}$$

by separation of the variables (see Exercise 1.2). We obtain the equation

$$x - \ln|x+1| = \tfrac{1}{2}t^2 + C, \tag{1.9}$$

C a constant, for the family of solution curves as well as the solution $x(t) \equiv -1$. However, to sketch the solution curves from (1.9) is less straightforward than to use (1.8) itself.

The above discussion has introduced two important ideas:

1. that two different differential equations can have solutions that exhibit the same qualitative behaviour; and
2. that the qualitative behaviour of solutions is determined by $X(t, x)$.

We will now put these two ideas together and illustrate the qualitative approach to differential equations for the special case of equations of the form $\dot{x} = X(x)$. We shall see that such equations can be classified into qualitatively equivalent types.

1.2 Autonomous equations

1.2.1 *Solution curves and the phase portrait*

A differential equation of the form

$$\dot{x} = X(x), \qquad x \in S \subseteq \mathsf{R}, \qquad (D = \mathsf{R} \times S) \tag{1.10}$$

is said to be *autonomous*, because $\dot{x}$ is determined by x alone and so the solutions are, as it were, self-governing.

The solutions of autonomous equations have the following important property. If $\xi(t)$ is a solution of (1.10) with domain I and range $\xi(I)$ then $\eta(t) = \xi(t+C)$, for any real C, is also a solution with the same range, but with domain $\{t \mid t + C \in I\}$. This follows because

$$\dot{\eta}(t) = \dot{\xi}(t+C) = X(\xi(t+C)) = X(\eta(t)). \tag{1.11}$$

The solution curve $x = \xi(t)$ is obtained by *translating* the solution curve $x = \eta(t)$ by the amount C in the positive t-direction.

Furthermore if there exists a unique solution curve passing through each point of the strip $D' = \mathsf{R} \times \xi(I)$ then *all* solution curves on D' are translations of $x = \xi(t)$. The domain D is therefore divided into strips where the solution curves are all obtained by shifting a single curve in the t-direction (see Figs. 1.12–1.15). For example,

$$\dot{x} = x \tag{1.12}$$

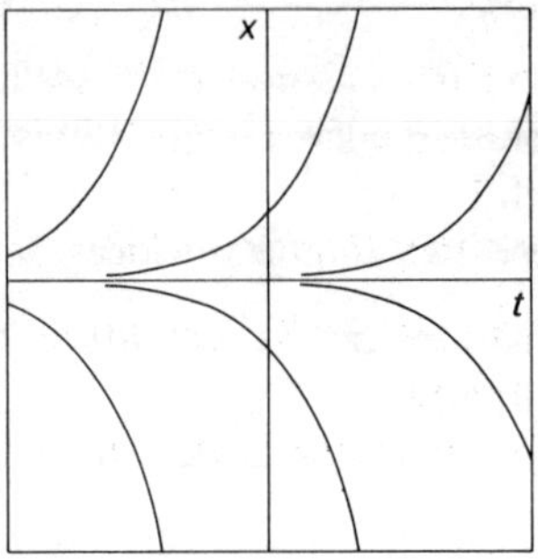

Fig. 1.12. $\dot{x} = x$: strips D' consist of the half planes $x < 0$ and $x > 0$.

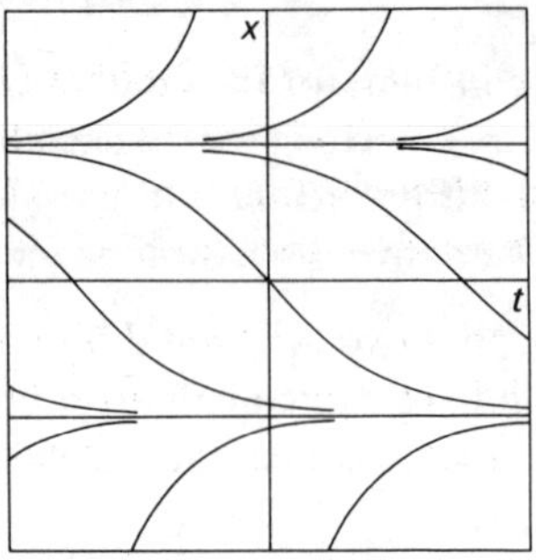

Fig. 1.13. $\dot{x} = \frac{1}{2}(x^2 - 1)$: strips $D' = \mathsf{R} \times \xi(I)$ with $\xi(I) = (-\infty, -1), (-1, 1)$ and $(1, \infty)$.

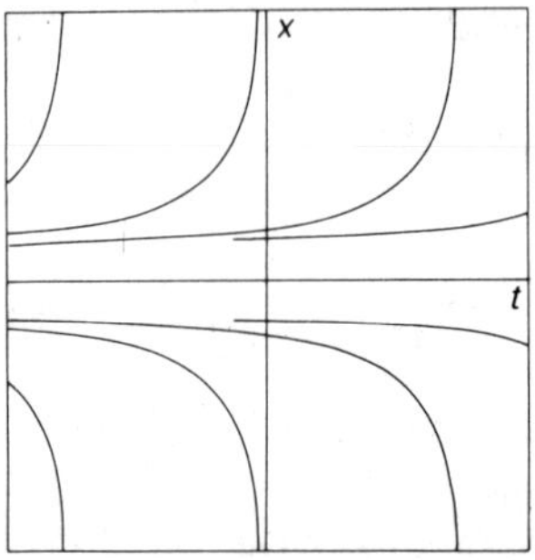

Fig. 1.14. Solution curves for $\dot{x} = x^3$.

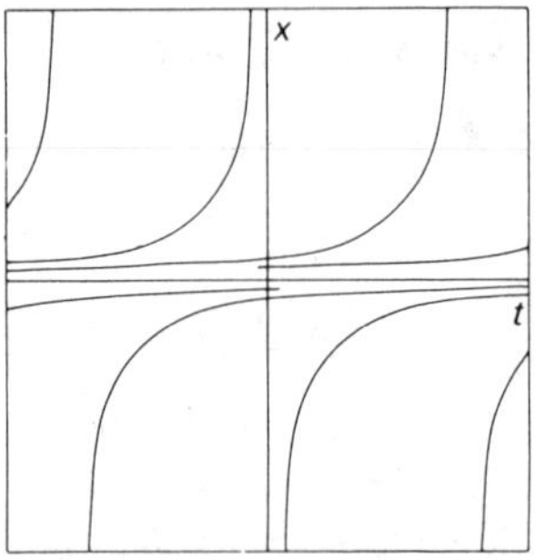

Fig. 1.15. Solution curves for $\dot{x} = x^2$.

has solutions:

$$\xi(t) = \mathrm{e}^t, \quad I = \mathsf{R}, \quad \xi(I) = (0, \infty) \tag{1.13}$$

$$\xi(t) \equiv 0, \quad I = \mathsf{R}, \quad \xi(I) = \{0\} \tag{1.14}$$

$$\xi(t) = -\mathrm{e}^t, \quad I = \mathsf{R}, \quad \xi(I) = (-\infty, 0). \tag{1.15}$$

All the solution curves in the strip D' defined by $x \in (0, \infty)$, $t \in \mathsf{R}$ are translations of e^t. Similarly, those in $D' = \{(t, x) \,|\, x \in (-\infty, 0), t \in \mathsf{R}\}$ are translations of e^{-t}.

For families of solution curves related by translations in t, the qualitative behaviour of the family of solutions is determined by that of any individual member. The qualitative behaviour of such a sample curve is determined by $X(x)$. When $X(x) \neq 0$, then the solution is either increasing or decreasing; when $X(c) = 0$ there is a solution $x(t) \equiv c$.

This information can be represented on the x-line rather than the t, x-plane. If $X(x) \neq 0$ for $x \in (a, b)$ then the interval is labelled with an arrow showing the sense in which x is changing. When $X(c) = 0$, the solution $x(t) \equiv c$ is represented by the point $x = c$. These solutions are called *fixed points* of the equation because x remains at c for all t. This geometrical representation of the qualitative behaviour of $\dot{x} = X(x)$ is called its *phase portrait*. Some examples of phase portraits are shown, in relation to X, in Figs. 1.16–1.19. The corresponding families of solution curves are given in Figs. 1.12–1.15.

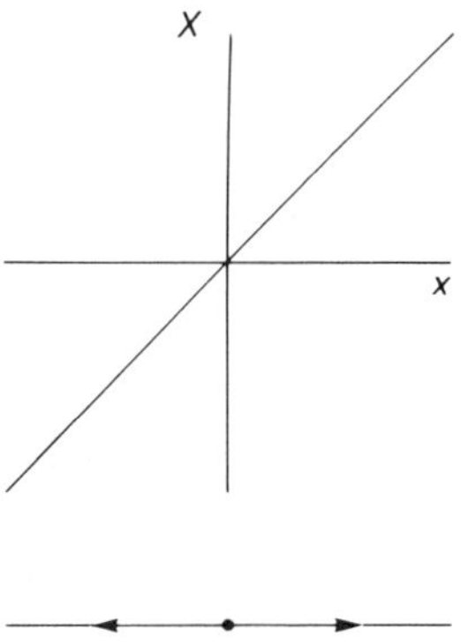

Fig. 1.16. $\dot{x} = x$, $x = 0$ is a fixed point.

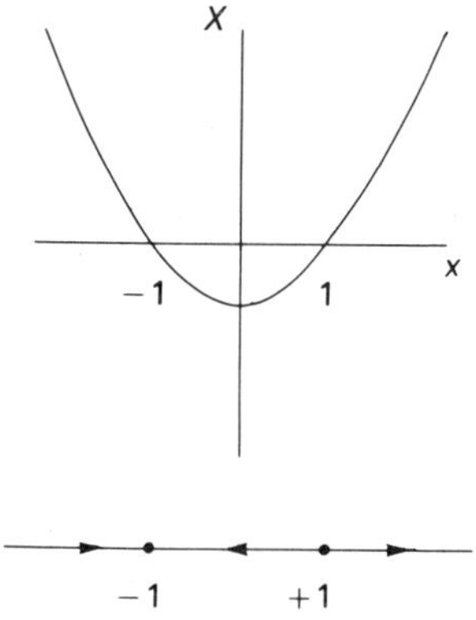

Fig. 1.17. $\dot{x} = \frac{1}{2}(x^2 - 1)$, $x = \pm 1$ are fixed points.

If x is not stationary it must either be increasing or decreasing. Thus for a given finite number of fixed points there can only be a finite number of 'different' phase portraits. By 'different', we mean with distinct assignments of where x is increasing or decreasing. For example, consider a single fixed point at $x = c$ (see Fig. 1.20). For

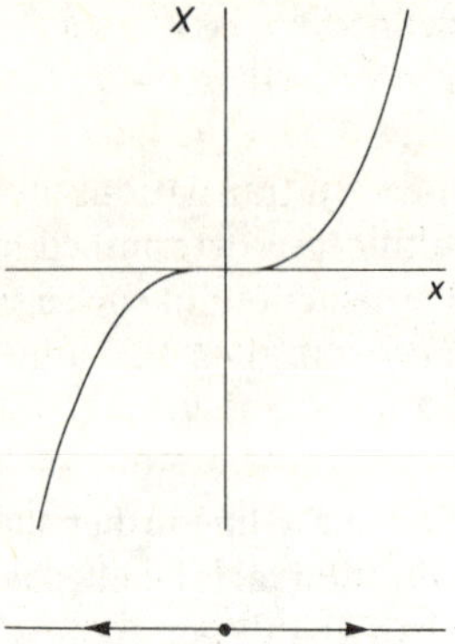

Fig. 1.18. $\dot{x} = x^3$, $x = 0$ is a fixed point.

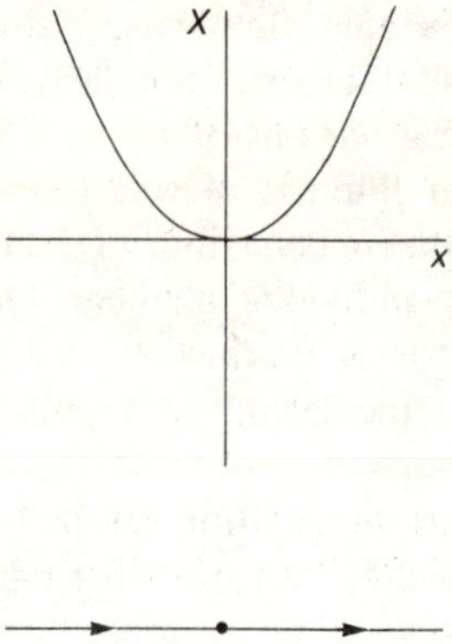

Fig. 1.19. $\dot{x} = x^2$, $x = 0$ is a fixed point.

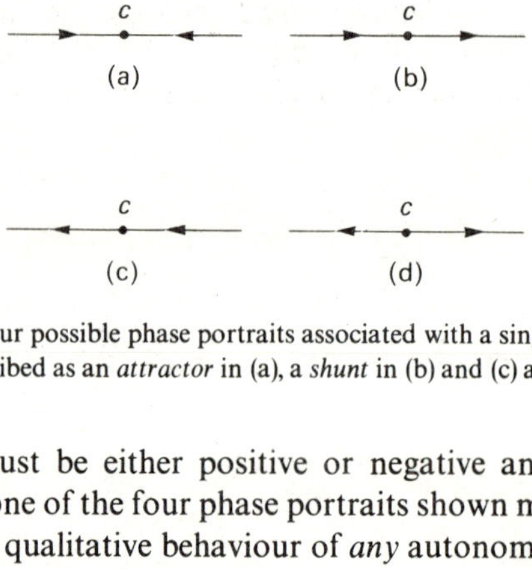

Fig. 1.20. The four possible phase portraits associated with a single fixed point. The fixed point is described as an *attractor* in (a), a *shunt* in (b) and (c) and a *repellor* in (d).

$x < c$, $X(x)$ must be either positive or negative and similarly for $x > c$. Hence, one of the four phase portraits shown must occur. This means that the qualitative behaviour of *any* autonomous differential equation with one fixed point *must* correspond to one of the phase portraits in Fig. 1.20 for some value of c. For example, $\dot{x} = x$, $\dot{x} = x^3$, $\dot{x} = x - a$, $\dot{x} = (x - a)^3$, $\dot{x} = \sinh x$, $\dot{x} = \sinh(x - a)$ all correspond to Fig. 1.20(d) for $c = 0$ or a. Of course, two different equations, each having one fixed point, that correspond to the *same* phase portrait in Fig. 1.20(d) for $c = 0$ or a. Of course, two different equations, each differential equations are *qualitatively equivalent*.

Now observe that the argument leading to Fig. 1.20 holds equally well if the fixed point at $x = c$ is one of many in a phase portrait. In other words, the qualitative behaviour of x in the neighbourhood of *any* fixed point *must* be one of those illustrated in Fig. 1.20(a)–(d). We say that this behaviour determines the *nature of the fixed point* and use the terminology defined in the caption to Fig. 1.20 to describe this.

This is an important step because it implies that the phase portrait of any autonomous equation is determined completely by the nature of its fixed points. We can make the following definition.

Definition 1.2.1. Two differential equations of the form $\dot{x} = X(x)$ are qualitatively equivalent if they have the same number of fixed points, of the same nature, arranged in the same order along the phase line.

For example, $\dot{x} = (x+2)(x+1)$ is qualitatively equivalent to $\dot{x} = \frac{1}{2}(x^2-1)$. Both have two fixed points, one attractor and one repellor, with the attractor occurring at the smaller value of x. The equation $\dot{x} = -(x+2)(x+1)$ is *not* qualitatively equivalent to $\dot{x} = \frac{1}{2}(x^2-1)$ because the attractor and repellor occur in the reverse order.

Example 1.2.1. Arrange the following differential equations in qualitatively equivalent groups:

(1) $\dot{x} = \cosh x$;	(2) $\dot{x} = (x-a)^2$;	(3) $\dot{x} = \sin x$;
(4) $\dot{x} = \cos x - 1$;	(5) $\dot{x} = \cosh x - 1$;	(6) $\dot{x} = \sin 2x$;
(7) $\dot{x} = e^x$;	(8) $\dot{x} = \sinh^2(x-b)$.	

Solution. Equations (1) and (7) have no fixed points; in both cases $X(x) > 0$ for all x. They both have the phase portrait shown in Fig. 1.21(a) and are therefore qualitatively equivalent.

Equation (2) has a single shunt at $x = a$ with $X(x) \geqslant 0$ for all x. Another equation with a single fixed point is (5); this has the same kind

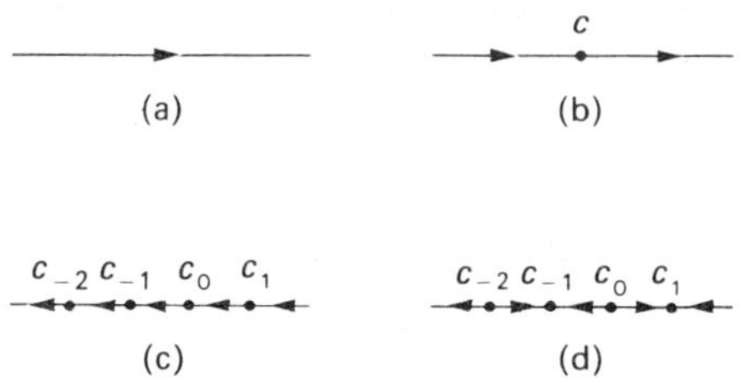

Fig. 1.21. Phase portraits for qualitatively equivalent groups in Example 1.2.1. In (b) $c = a$ for (2), 0 for (5) and b for (8). In (c) $c_n = 2n\pi$ for (4) and in (d) $c_n = n\pi$ for (3) and $n\pi/2$ for (6).

of shunt at $x = 0$. Equation (8) also has a single shunt, at $x = b$; again $X(x) \geqslant 0$ for all x. These equations form a second qualitatively equivalent group with the phase portrait shown in Fig. 1.21(b).

The remaining equations, (3), (4) and (6), all have infinitely many fixed points: (3) at $x = n\pi$, (4) at $x = 2n\pi$ and (6) at $x = n\pi/2$. However, equation (4) has $X(x) \leqslant 0$ for all x, so that every fixed point is a shunt, whereas (3) and (6) have alternating attractors and repellors (see Fig. 1.21(c) and (d)). Thus only (3) and (6) are qualitatively equivalent. □

Example 1.2.1 draws attention to the fact that 'qualitative equivalence' is an equivalence relation on the set of all autonomous differential equations. We can consequently divide this set into disjoint classes according to their qualitative behaviour. However, if we only demand uniqueness of solutions then there are infinitely many distinct qualitative classes. This follows because arbitrarily many fixed points can occur.

If we place other limitations on X, there may only be a finite number of classes. For example, suppose we require X to be *linear*, i.e. $X(x) = ax$, a real. For any $a \neq 0$, there is only a single fixed point at $x = 0$. If $a > 0$, this is a repellor and if $a < 0$ it is an attractor. For the special case $a = 0$ every point of the phase portrait is a fixed point. Thus, $\{\dot{x} = X(x) \,|\, X(x) = ax\}$ can be divided into three classes according to qualitative behaviour.

For non-linear X, each isolated fixed point can only be one of the four possibilities shown in Fig. 1.20. Thus, although there are infinitely many distinct phase portraits, they contain, at most, four distinct types of fixed point. This limitation arises because $\dot{x} = X(x)$ is a differential equation for the *single* real variable x. There is consequently a one-dimensional phase portrait with x either increasing or decreasing at a non-stationary point.

1.2.2 *Phase portraits and dynamics*

In applications the differential equation $\dot{x} = X(x)$ models the *time* dependence of a property, x, of some physical system. We say that the *state* of the system is specified by x. For example, the equation

$$\dot{p} = ap; \qquad p, a > 0 \tag{1.16}$$

models the growth of the population, p, of an isolated species. Within this model, the state of the species at time t is given by the number of individuals, $p(t)$, living at that time. Another example is Newton's law of cooling. The temperature, T, of a body cooling in a draught of temperature τ is given by

$$\dot{T} = -a(T-\tau), \qquad a > 0. \tag{1.17}$$

Here the state of the body is taken to be determined by its temperature.

We can represent the state $x(t_0)$ of a model at any time t_0 by a point on the phase line of $\dot{x} = X(x)$. As time increases, the state changes and the phase point representing it moves along the line with velocity $\dot{x} = X(x)$. Thus, the dynamics of the physical system are represented by the motion of a phase point on the phase line.

The phase portrait records only the *direction* of the velocity of the phase point and therefore represents the dynamics in a qualitative way. Such qualitative information can be helpful when constructing models. For example, consider the model (1.16) of an isolated population. Observe that $\dot{p} > 0$ for *all* $p > 0$, the phase portrait, in Fig. 1.22(a), shows that the population increases indefinitely. This feature is clearly unrealistic; the environment in which the species live must have limits and could not support an ever-increasing population.

Fig. 1.22. Phase portraits for (a) $\dot{p} = ap$ and (b) $\dot{p} = p(a-bp)$, $p_e = a/b$. In both cases, we are interested only in the behaviour for non-negative populations ($p \geqslant 0$). The equation in (b) is known as the *logistic law* of population growth.

Let us suppose that the environment can support a population p_e, then how could (1.16) be modified to take account of this? Obviously, the indefinite increase of p should be interrupted. One possibility is to introduce an attractor at p_e as shown in Fig. 1.22(b). This means that populations greater than p_e decline, while populations less than p_e increase. Finally, equilibrium is reached at $p = p_e$. The fixed point at $p = p_e$, as well as $p = 0$, requires a non-linear $X(p)$ in (1.16). The form

$$\dot{p} = p(a-bp) \tag{1.18}$$

has the advantage of reducing to (1.16) when $b = 0$; otherwise $p_e = a/b$.

Of course, models of physical systems frequently involve more than a single state variable. If we are to be able to use qualitative ideas in modelling these systems, then we must examine autonomous equations involving more than one variable.

1.3 Autonomous systems in the plane

Consider the differential equation

$$\dot{\mathbf{x}} \equiv \frac{d\mathbf{x}}{dt} = \mathbf{X}(\mathbf{x}) \tag{1.19}$$

where $\mathbf{x} = (x_1, x_2)$ is a vector in R^2. This equation is equivalent to the system of two coupled equations

$$\dot{x}_1 = X_1(x_1, x_2), \qquad \dot{x}_2 = X_2(x_1, x_2), \tag{1.20}$$

where $\mathbf{X}(\mathbf{x}) = (X_1(x_1, x_2),\ X_2(x_1, x_2))$, because $\dot{\mathbf{x}} = (\dot{x}_1, \dot{x}_2)$. A solution to (1.19) consists of a pair of functions $(x_1(t), x_2(t))$, $t \in I \subseteq \mathsf{R}$, which satisfy (1.20). In general, both $x_1(t)$ and $x_2(t)$ involve an arbitrary constant so that there is a two-parameter family of solutions.

The qualitative behaviour of this family is determined by how x_1 and x_2 behave as t increases. Instead of simply indicating whether x is increasing or decreasing on the phase line, we must indicate how $\mathbf{x}$ behaves in the *phase plane*. The phase portrait is therefore a two-dimensional figure and the qualitative behaviour is represented by a family of curves, directed with increasing t, known as *trajectories* or *orbits*.

To examine qualitative behaviour in the plane, we begin (as in Section 1.2) by looking at any *fixed points* of (1.19). These are solutions of the form $\mathbf{x}(t) \equiv \mathbf{c} = (c_1, c_2)$ which arise when

$$X_1(c_1, c_2) = 0 \quad \text{and} \quad X_2(c_1, c_2) = 0. \tag{1.21}$$

Their corresponding trajectory is the point (c_1, c_2) in the phase plane. In Section 1.2, the 'nature' of the fixed points determined the phase portrait. Let us consider some examples of isolated fixed points in the plane, with a view to determining their nature. Figures 1.23–1.32 show a small selection of the possibilities.

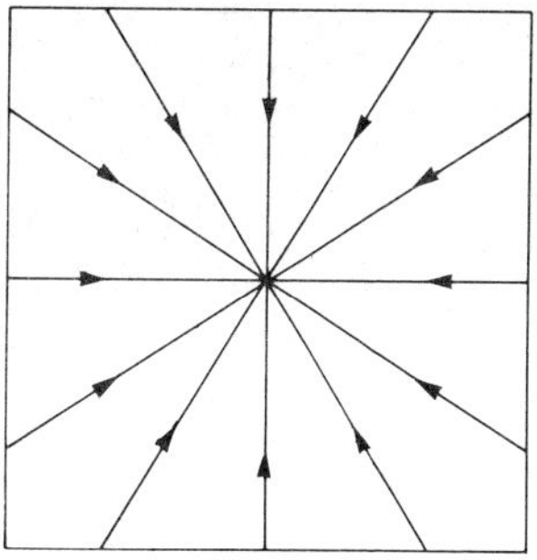

Fig. 1.23. $\dot{x}_1 = -x_1, \dot{x}_2 = -x_2$.

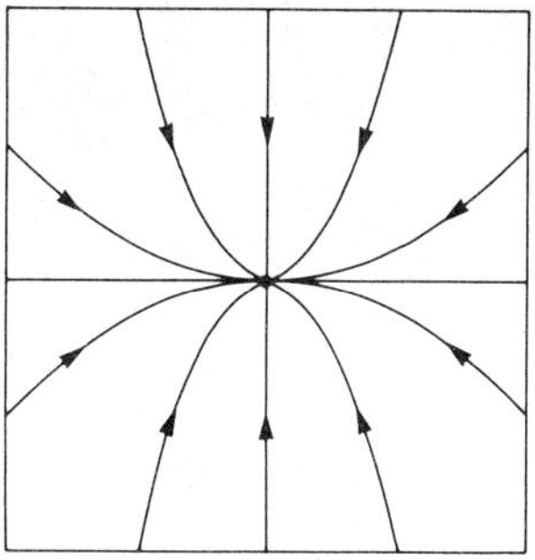

Fig. 1.24. $\dot{x}_1 = -x_1, \dot{x}_2 = -2x_2$.

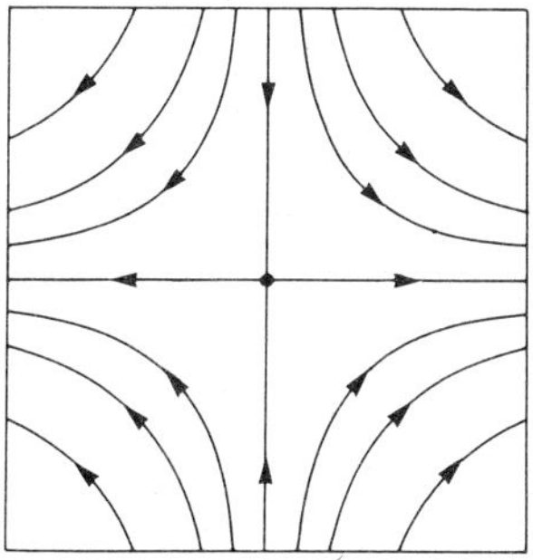

Fig. 1.25. $\dot{x}_1 = x_1, \dot{x}_2 = -x_2$.

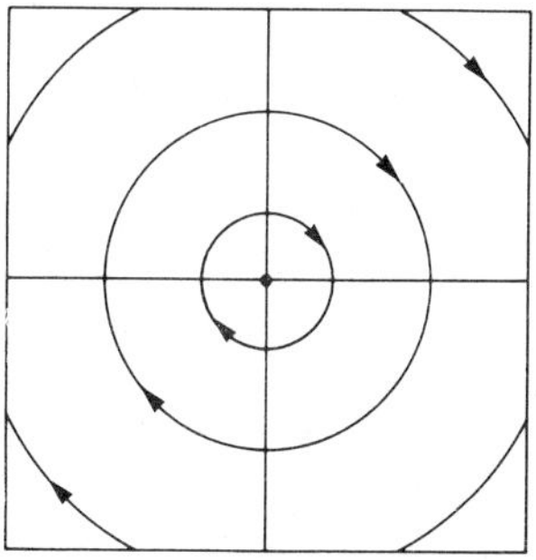

Fig. 1.26. $\dot{x}_1 = x_2, \dot{x}_2 = -x_1$.

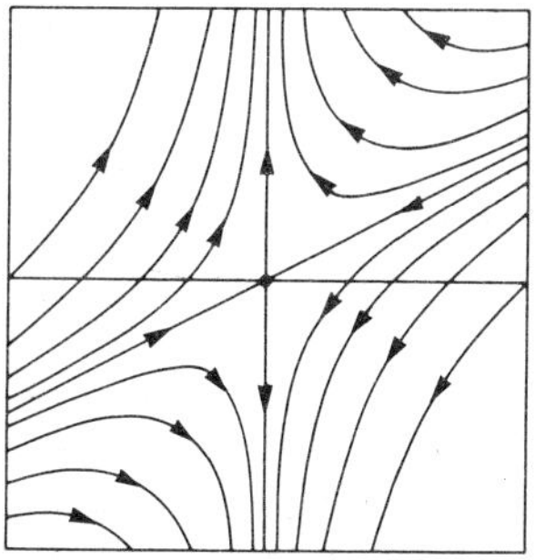

Fig. 1.27. $\dot{x}_1 = -x_1, \dot{x}_2 = -x_1 + x_2$.

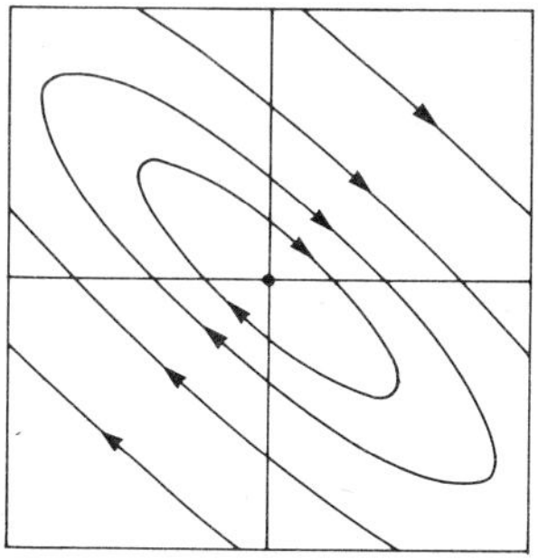

Fig. 1.28. $\dot{x}_1 = 3x_1 + 4x_2$,
$\dot{x}_2 = -3x_1 - 3x_2$.

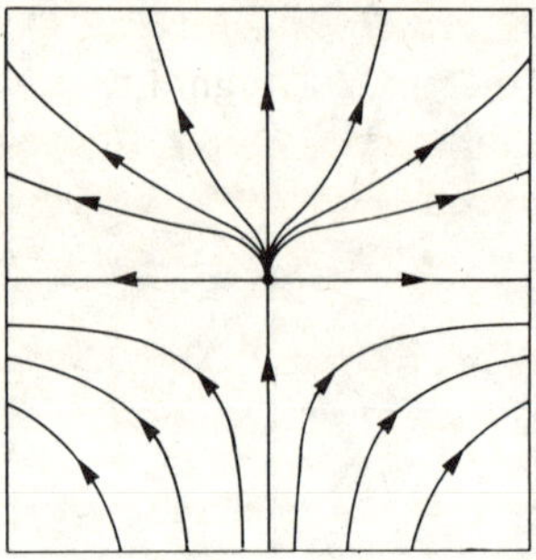

Fig. 1.29. $\dot{x}_1 = x_1, \dot{x}_2 = x_2^2$.

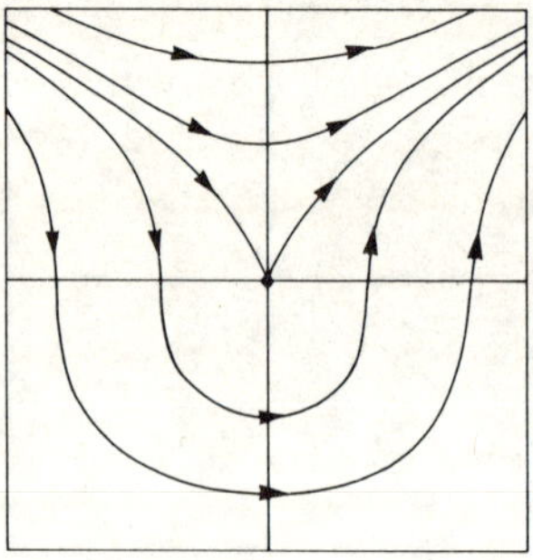

Fig. 1.30. $\dot{x}_1 = x_2^2, \dot{x}_2 = x_1$.

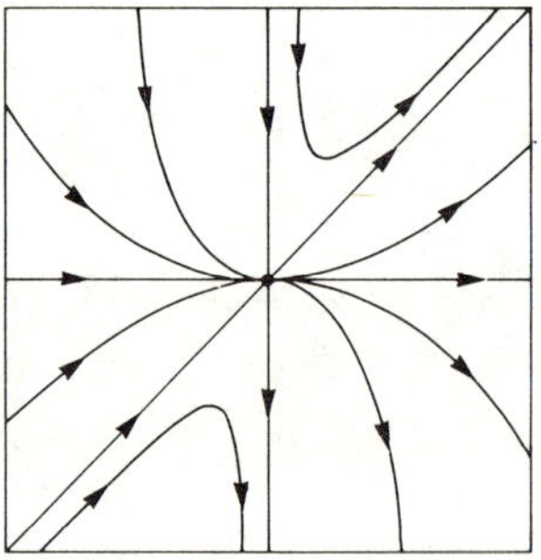

Fig. 1.31. $\dot{x}_1 = x_1^2, \dot{x}_2 = x_2(2x_1 - x_2)$.

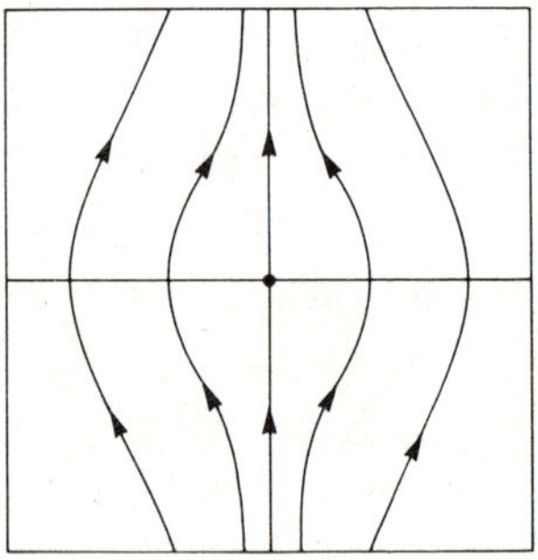

Fig. 1.32. $\dot{x}_1 = -x_1 x_2, \dot{x}_2 = x_1^2 + x_2^2$.

Consider Fig. 1.23; the system

$$\dot{x}_1 = -x_1, \qquad \dot{x}_2 = -x_2 \tag{1.22}$$

has a fixed point at (0, 0) and solutions

$$x_1(t) = C_1 e^{-t}, \qquad x_2(t) = C_2 e^{-t} \tag{1.23}$$

where C_1, C_2 are real constants. Clearly, every member of the family (1.23) satisfies

$$x_2(t) = K x_1(t), \tag{1.24}$$

where $K = C_2/C_1$, for every t. Thus every member is associated with a radial straight line in the x_1x_2-plane. Equation (1.23) shows that, for any choice of C_1 and C_2, $|x_1(t)|$ and $|x_2(t)|$ decrease as t increases and approach zero as $t \to \infty$. This is indicated by the direction of the arrow on the trajectory; i.e. if $(x_1(t), x_2(t))$ represents the value of $\mathbf{x}$ at t then this *phase point* will move along the radial straight line towards the origin as t increases. The directed straight line is sufficient to describe this qualitative behaviour. In Fig. 1.24,

$$x_2 = Kx_1^2,$$

which alters the shape of the trajectories. However, they are still all directed towards the fixed point at the origin.

Figure 1.25 shows another possibility: here as t increases, $|x_1(t)|$ decreases but $|x_2(t)|$ increases. In fact, $\dot{x}_1 = -x_1$, $\dot{x}_2 = x_2$ has solutions

$$x_1(t) = C_1 e^{-t}, \qquad x_2(t) = C_2 e^{t}, \tag{1.25}$$

C_1, C_2 real; so that

$$x_2 = Kx_1^{-1}, \tag{1.26}$$

with $K = C_1 C_2$. In this case, only two special trajectories approach the fixed point at (0, 0), the remainder all turn away sooner or later and $|x_2| \to \infty$ as $|x_1| \to 0$. This qualitative behaviour is obviously quite different from that in Figs. 1.23 and 1.24.

In Fig. 1.26 the trajectories close on themselves so that the same set of points in the phase plane recur time and time again as t increases. In Section 1.4, we show that the system

$$\dot{x}_1 = x_2, \qquad \dot{x}_2 = -x_1 \tag{1.27}$$

has solutions

$$x_1(t) = C_1 \cos(-t + C_2), \qquad x_2(t) = C_1 \sin(-t + C_2). \tag{1.28}$$

It follows that

$$x_1^2 + x_2^2 = C_1^2 \tag{1.29}$$

and the trajectories are a family of concentric circles centred on the fixed point at (0, 0). This obviously corresponds to yet another kind of qualitative behaviour. The fact that $x_1(t)$ and $x_2(t)$ are periodic with the same period is reflected in the closed trajectories.

These examples show that qualitatively different solutions, $(x_1(t), x_2(t))$, lead to trajectories with different geometrical properties. The problem of recognizing different types of fixed point becomes one of recognizing 'distinct' geometrical configurations of trajectories. As in Section 1.2, we must decide what we mean by 'distinct' and there is an element of choice in the criteria that we set.

For example, in Figs. 1.23 and 1.24 all the trajectories are directed toward the origin. It would be reasonable to argue that this is the dominant qualitative feature and that the differences in shape of the trajectories are unimportant. We would then say that the nature of the fixed point at (0, 0) was the same in both cases. Of course, its nature would be completely changed if we replaced $\dot{x}_1$ by $-\dot{x}_1$ and $\dot{x}_2$ by $-\dot{x}_2$. Under these circumstances all trajectories would be directed away from the origin (see Exercise 1.22) corresponding to quite different qualitative behaviour of the solutions.

Let us compare Figs. 1.25 and 1.27. Are the fixed points of the same nature? In both cases $|x_1(t)|$ tends to zero while $|x_2(t)|$ becomes infinite and only two special trajectories approach the fixed point itself. Yes, we would argue, they are the same. If the orientation of the trajectories is reversed in these examples is the nature of the fixed point changed as in our previous example? Orientation reversal would mean that the roles of x_1 and x_2 were interchanged. However, the features which distinguish Figs. 1.25 and 1.27 from the remaining ten diagrams still persist and we conclude that the nature of the fixed point does not change. Similarly, we would say that Figs. 1.26, 1.28 and their counterparts with orientation reversed all had the same kind of fixed point at the origin.

The intuitive approach used in discussing the above examples is sufficient for us to recognize that seven distinct types of fixed point are illustrated in Figs. 1.23–1.32. In fact, there are infinitely many qualitatively different planar phase portraits containing a single fixed point. Some examples of phase portraits with more than one isolated fixed point are shown in Figs. 1.33–1.36. As can be seen, it is not difficult to produce complicated families of trajectories.

1.4 Construction of phase portraits in the plane

Methods of obtaining information about the trajectories associated with systems like

$$\dot{x}_1 = X_1(x_1, x_2); \qquad \dot{x}_2 = X_2(x_1, x_2), \qquad (x_1, x_2) \in S \subseteq \mathsf{R}^2$$

are straightforward extensions of ideas used to obtain solutions in Section 1.1.

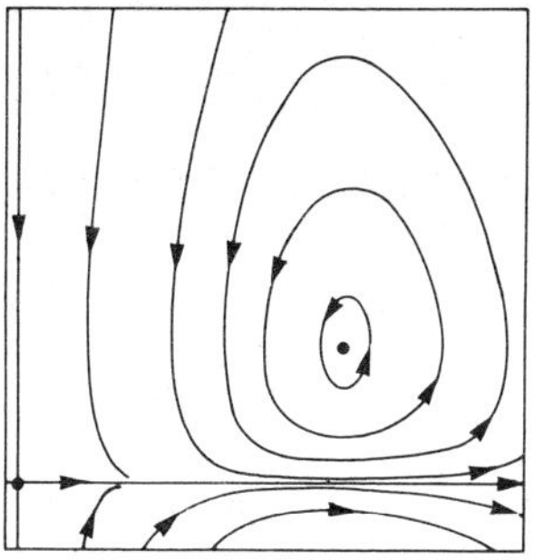

Fig. 1.33. Phase portrait for $\dot{x}_1 = x_1(a - bx_2)$, $\dot{x}_2 = -x_2(c - dx_1)$ with $a, b, c, d > 0$. There are fixed points at $(0, 0)$ and $(c/d, a/b)$ in the phase plane.

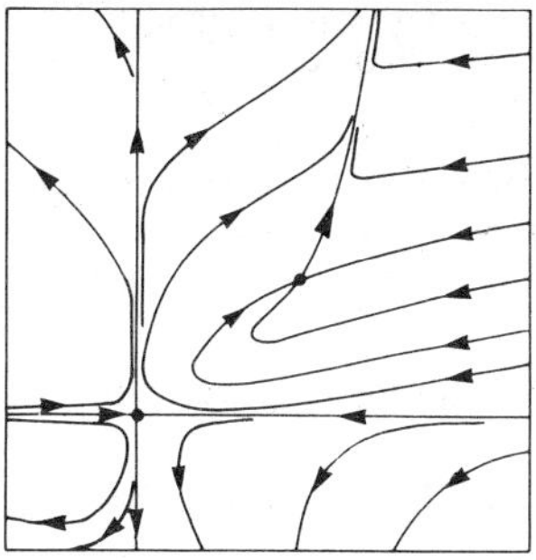

Fig. 1.34. Phase portrait for $\dot{x}_1 = -x_1[1 - 3x_2^{2/3}(1 - x_1)/(1 + x_1)]$, $\dot{x}_2 = x_2 - 3x_1x_2^{2/3}/(1 + x_1)$; $x_1, x_2 \geqslant 0$. Fixed points occur at $(0, 0)$ and $(\frac{1}{2}, 1)$.

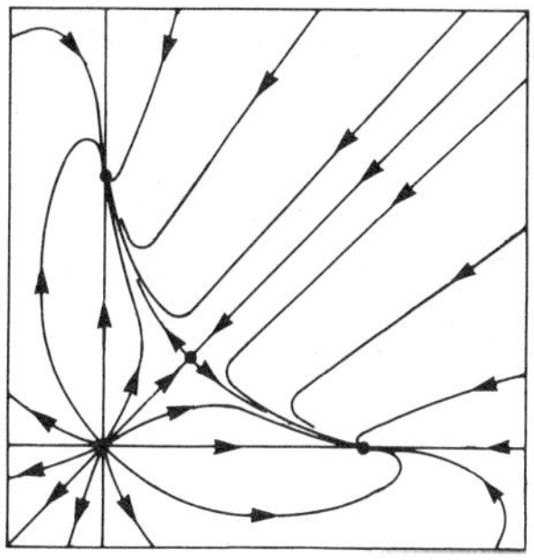

Fig. 1.35. Phase portrait for $\dot{x}_1 = (2 - x_1 - 2x_2)x_1$, $\dot{x}_2 = (2 - 2x_1 - x_2)x_2$. Fixed points occur at $(0,0)$, $(2,0)$, $(0,2)$ and $(\frac{2}{3}, \frac{2}{3})$.

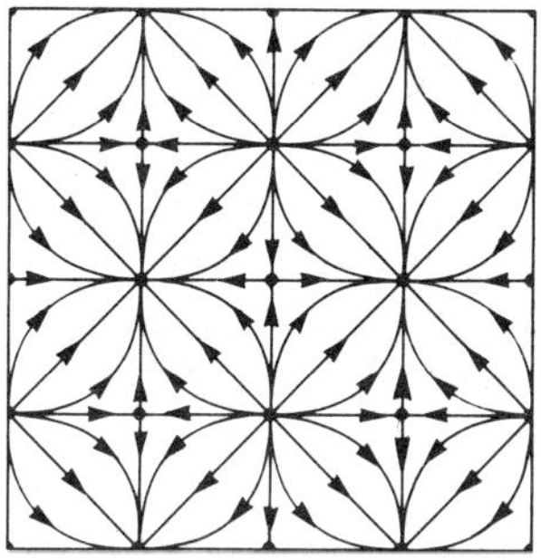

Fig. 1.36. Phase portrait for $\dot{x}_1 = \sin x_1$, $\dot{x}_2 = -\sin x_2$. Fixed points occur at $(n\pi, m\pi)$; n, m integers.

1.4.1 *Use of calculus*

In Section 1.3, we were able to obtain the trajectories shown in Figs. 1.23–1.25 by solving the system equations separately. This was possible because $\dot{x}_1$ depended only on x_1 and $\dot{x}_2$ only on x_2. A system in which each equation contains one, and only one, variable is said to be *decoupled*. This is not usually the case and if calculus is to be used in this way, new variables must be found which bring about this isolation of the variables.

Example 1.4.1. Consider the system

$$\dot{x}_1 = x_2, \qquad \dot{x}_2 = -x_1 \tag{1.30}$$

whose phase portrait appears in Fig. 1.26. Use plane polar coordinates (r, θ) such that

$$x_1 = r\cos\theta, \qquad x_2 = r\sin\theta \tag{1.31}$$

to re-express (1.30) and hence obtain x_1 and x_2 as functions of t.

Solution. We establish differential equations for r and θ by observing that

$$r^2 = x_1^2 + x_2^2 \quad \text{and} \quad \tan\theta = x_2/x_1, \quad x_1 \neq 0, \tag{1.32}$$

and differentiating with respect to t. We find

$$2r\dot{r} = 2x_1\dot{x}_1 + 2x_2\dot{x}_2 \quad \text{and} \quad \sec^2\theta\dot{\theta} = x_1^{-2}(\dot{x}_2 x_1 - \dot{x}_1 x_2). \tag{1.33}$$

On substituting for $\dot{x}_1$ and $\dot{x}_2$ from (1.30) we conclude that

$$\dot{r} = 0 \quad \text{and} \quad \dot{\theta} = -1. \tag{1.34}$$

These equations can be solved separately to obtain $r(t) \equiv C_1$ and $\theta(t) = -t + C_2$, where C_1 and C_2 are real constants. Finally, we have

$$x_1(t) = C_1\cos(-t + C_2) \quad \text{and} \quad x_2(t) = C_1\sin(-t + C_2) \tag{1.35}$$

from (1.31). □

Sometimes solutions can be found when only one of the variables is isolated in one of the equations. Such systems are said to be *partially decoupled.*

Example 1.4.2. Find solutions to the system

$$\dot{x}_1 = x_1, \qquad \dot{x}_2 = x_1 + x_2 \tag{1.36}$$

and hence construct its phase portrait.

Solution. The system (1.36) can be solved without introducing new variables. The first equation has solutions

$$x_1(t) = C_1 e^t, \qquad C_1 \text{ real}. \tag{1.37}$$

Substitution in the second gives

$$\dot{x}_2 = x_2 + C_1 e^t \tag{1.38}$$

which in turn has solutions

$$x_2(t) = \mathrm{e}^t(C_1 t + C_2) \tag{1.39}$$

(see Exercise 1.1).

To construct the phase portrait examine (1.37) and (1.39) and note:

(a) For $C_1 = K > 0$, $x_1(t)$ strictly increases through all positive values as t increases through $(-\infty, \infty)$.

(b) For $C_1 = K > 0$, $x_2(t) \to 0$ as $t \to -\infty$; $x_2(t) < 0$ for $C_1 t + C_2 < 0$; $x_2(t) = 0$ for $C_1 t + C_2 = 0$ and $x_2(t) \to \infty$ as $t \to \infty$.

(c) For $C_1 = 0$, we obtain the solution

$$x_1(t) \equiv 0, \qquad x_2(t) = C_2 \mathrm{e}^t. \tag{1.40}$$

(d) For $C_1 = -K < 0$, both $x_1(t)$ and $x_2(t)$ assume precisely minus one times the values on a trajectory in (a) and (b) above.

Notice also that the turning point in x_2, implied in (b), is given more precisely in (1.36) where $\dot{x}_2 = 0$ when $x_2 = -x_1$. Furthermore, the symmetry described in (d) is also apparent in (1.36) because this system is invariant under the transformation $x_1 \to -x_1$, $x_2 \to -x_2$ (see Exercise 1.26). Finally, we sketch the phase portrait in Fig. 1.37. □

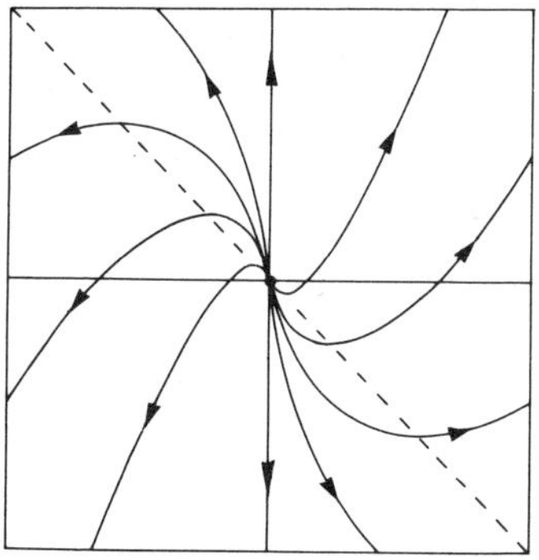

Fig. 1.37. Phase portrait for the system $\dot{x}_1 = x_1$, $\dot{x}_2 = x_1 + x_2$. The dashed line is $x_2 = -x_1$ where $\dot{x}_2 = 0$. The origin is a fixed point.

In Section 1.3, we sometimes found it advantageous to eliminate t between $x_1(t)$ and $x_2(t)$ and obtain a non-parametric form for the trajectories. These equations can often be found directly by solving

$$\frac{dx_2}{dx_1} = \frac{\dot{x}_2}{\dot{x}_1} = \frac{X_2(x_1, x_2)}{X_1(x_1, x_2)}. \tag{1.41}$$

Example 1.4.3. Consider the system

$$\dot{x}_1 = x_2, \qquad \dot{x}_2 = x_1. \tag{1.42}$$

Use dx_2/dx_1 to determine the nature of the trajectories and hence construct the phase portrait for the system.

Solution. For the system (1.42) we have

$$\frac{dx_2}{dx_1} = \frac{x_1}{x_2}, \quad x_2 \neq 0, \tag{1.43}$$

which has solutions satisfying

$$x_1^2 - x_2^2 = C, \tag{1.44}$$

C real. This family of hyperbolae is easily sketched and their orientation as trajectories is given by (1.42). For example, note that both $x_1, x_2 > 0$ implies that both $x_1(t)$ and $x_2(t)$ increase with t. This provides directions for all the trajectories in $x_2 > -x_1$. Similarly, $x_1, x_2 < 0$ means $\dot{x}_1, \dot{x}_2 < 0$ and gives orientation to the trajectories in $x_2 < -x_1$. Thus the phase portrait is as shown in Fig. 1.38. □

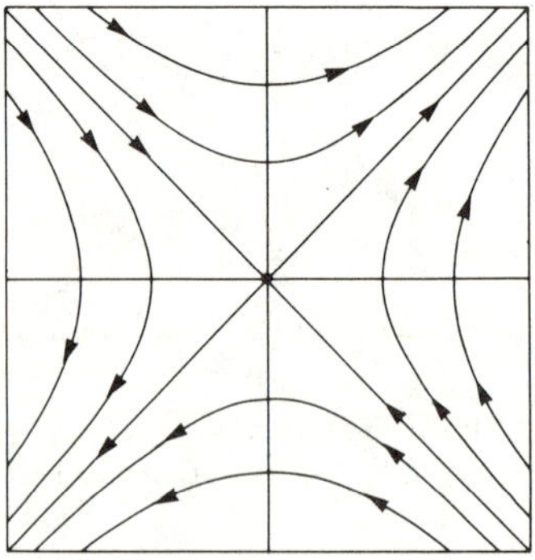

Fig. 1.38. Phase portrait for $\dot{x}_1 = x_2$, $\dot{x}_2 = x_1$. The origin is a fixed point and the trajectories are hyperbolae.

1.4.2 *Isoclines*

As in Section 1.1, it may be necessary to construct a phase portrait when calculus fails to give tractable solutions. This can be done by extending the method of isoclines to the plane. The vector valued function or *vector field* $\mathbf{X}: S \to \mathsf{R}^2$ now gives $\dot{\mathbf{x}}$ at each point of the plane where $\mathbf{X}$ is defined. For qualitative purposes it is usually sufficient to record the direction of $\mathbf{X}(\mathbf{x})$. This is constant on the isoclines of (1.41). The zeroes and divergences of dx_2/dx_1 are of particular interest and such lines will be referred to as the '$\dot{x}_2 = 0$' or '$\dot{x}_1 = 0$'-isoclines respectively.

If a unique solution $\mathbf{x}(t)$ of

$$\dot{\mathbf{x}} = \mathbf{X}(\mathbf{x}), \qquad \mathbf{x} \in S \quad \text{and with} \quad \mathbf{x}(t_0) = \mathbf{x}_0 \tag{1.45}$$

exists for any $\mathbf{x}_0 \in S$ and $t_0 \in \mathsf{R}$, then each point of S lies on one, and only one, trajectory. All of the examples presented thus far in Sections 1.3 and 1.4 have been of this kind. Furthermore, when non-uniqueness occurs, it is frequently confined to lines in the domain S. It is only in the neighbourhood of such lines that the behaviour of the trajectories is not immediately apparent from $\mathbf{X}(\mathbf{x})$.

Example 1.4.4. Obtain solutions to the system

$$\dot{x}_1 = 3x_1^{2/3}, \qquad \dot{x}_2 = 1 \tag{1.46}$$

and show that any point $(0, c)$, c real, lies on more than one solution. Interpret this result on the phase portrait for (1.46).

Solution. The system (1.46) is already decoupled. Calculus gives

$$x_1(t) = (t + C_1)^3 \quad \text{and} \quad x_2(t) = t + C_2 \tag{1.47}$$

as functions which satisfy the system equations. However, we know that $x_1(t) \equiv 0$ is a solution of $\dot{x}_1 = 3x_1^{2/3}$ and so there are solutions of the form

$$x_1(t) \equiv 0, \qquad x_2(t) = t + C_2. \tag{1.48}$$

If we let $C_2 = 0$ and $C_1 = -c$ in (1.47) then $(x_1(c), x_2(c)) = (0, c)$. Equally, if $C_2 = 0$ in (1.48) then $(x_1(c), x_2(c)) = (0, c)$, also. The solutions in (1.47) and (1.48) are distinct so $(0, c)$ lies on more than one solution.

The phase portrait obtained by representing (1.47) and (1.48) as trajectories is shown in Fig. 1.39. Notice the trajectories derived from

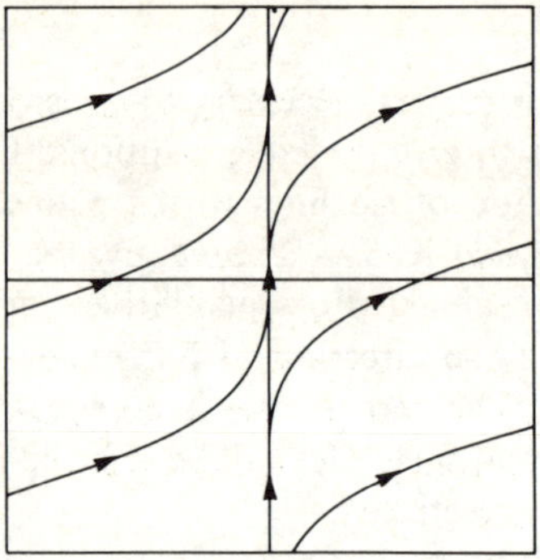

Fig. 1.39. Phase portrait for $\dot{x}_1 = 3x_1^{2/3}$, $\dot{x}_2 = 1$. There is no fixed point for this system and trajectories touch the x_2-axis.

(1.47) all touch the x_2-axis. This is itself a trajectory, oriented as shown, corresponding to (1.48). It is no longer clear what we should mean by a single trajectory because

$$x_1(t) = \begin{cases} (t-a)^3, & t < a \\ 0, & a \leqslant t \leqslant b, \\ (t-b)^3, & t > b, \end{cases} \qquad x_2 = t + C \tag{1.49}$$

also satisfies (1.46). □

In the remainder of this book we will be concerned only with systems that have unique solutions. However, (cf. Section 1.1.1) a useful sufficient condition for existence and uniqueness of solutions to $\dot{\mathbf{x}} = \mathbf{X}(\mathbf{x})$ is that $\mathbf{X}$ be continuously differentiable.

Example 1.4.5. Sketch the phase portrait for the system

$$\dot{x}_1 = x_1^2, \qquad \dot{x}_2 = x_2(2x_1 - x_2) \tag{1.50}$$

without using calculus to determine the trajectories.

Solution. Examine (1.50) and note the following features:

(a) there is a single fixed point at the origin of the x_1x_2-plane;

(b) the vector field $\mathbf{X}(\mathbf{x}) = (x_1^2, x_2(2x_1 - x_2))$, corresponding to (1.50), satisfies $\mathbf{X}(\mathbf{x}) = \mathbf{X}(-\mathbf{x})$. This means that the shape of the trajectories is invariant under the transformation $x_1 \to -x_1$ and $x_2 \to -x_2$ (see Exercise 1.26). We will, therefore, focus attention on the half plane $x_1 \geqslant 0$;

(c) the $\dot{x}_1 = 0$ isocline coincides with the x_2-axis and on it $\dot{x}_2 = -x_2^2 < 0$ for $x_2 \neq 0$. Hence, there is a trajectory coincident with the positive x_2-axis directed toward the origin and one coincident with the negative x_2-axis directed away from the origin.

Consider

$$\frac{dx_2}{dx_1} = \frac{x_2(2x_1 - x_2)}{x_1^2}, \qquad x_1 \neq 0. \tag{1.51}$$

The isocline of slope C is given by

$$x_2(2x_1 - x_2) = x_1^2 - (x_1 - x_2)^2 = Cx_1^2$$

or

$$x_2 = x_1(1 \pm (1 - C)^{1/2}), \qquad C \leqslant 1. \tag{1.52}$$

This equation allows us to obtain the slope of the trajectories on a selection of isoclines. For example, the slope C is given as:

(a) $C = 0$ on the lines $x_2 = 0$ and $x_2 = 2x_1$;
(b) $C = 1$ on $x_2 = x_1$;
(c) $C = \frac{1}{2}$ on $x_2 = (1 \pm 1/\sqrt{2})x_1$;
(d) $C = -3$ on $x_2 = -x_1$ and $x_2 = 3x_1$;
(e) $C = -2$ on $x_2 = (1 \pm \sqrt{3})x_1$.

The orientation of the trajectories is fixed by recognizing that $\dot{x}_1 > 0$ for $x_1 \neq 0$.

This information allows us to construct Fig. 1.40 and we can, using uniqueness, sketch the phase portrait shown in Fig. 1.41. □

1.5 Flows and evolution

The dynamic interpretation of the differential equation

$$\dot{x} = X(x), \qquad x \in S \subseteq \mathbb{R} \tag{1.53}$$

as the velocity of a point on the phase line (cf. Section 1.2.2), leads to a new way of looking at both the differential equation and its solutions. Equation (1.53) can be thought of as defining a *flow* of phase points along the phase line. The function X gives the velocity of the flow at each value of $x \in S$.

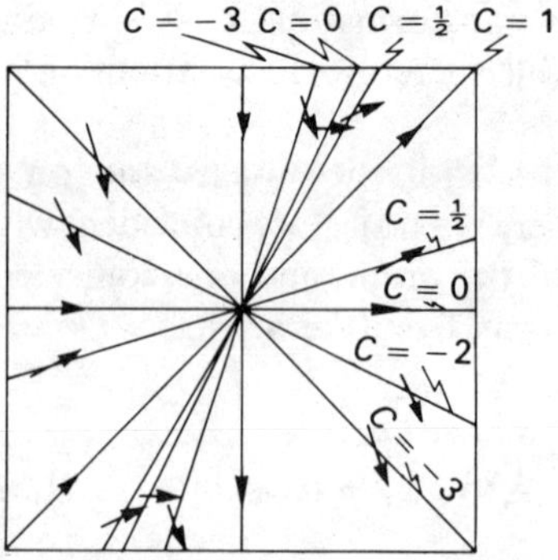

Fig. 1.40. Summary of information about $\dot{x}_1 = x_1^2$, $\dot{x}_2 = x_2(2x_1 - x_2)$. Note $C = 1$ on $x_2 = x_1$ means trajectories coincide with this line for $x_1 \neq 0$ as shown.

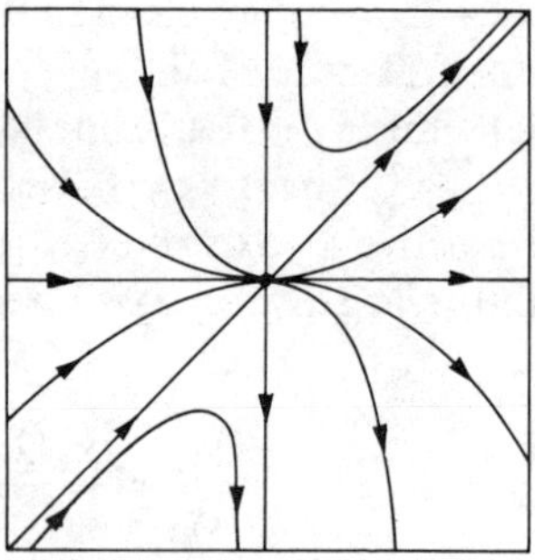

Fig. 1.41. Sketch of phase portrait for system (1.50). Notice curvature of trajectories between lines $x_2 = 0$ and $x_2 = x_1$ follows because C is increasing as shown in Fig. 1.40.

The solution, $x(t)$, of (1.53) which satisfies $x(t_0) = x_0$ gives the past ($t < t_0$) and future ($t > t_0$) positions, or *evolution*, of the phase point which is at x_0 when $t = t_0$. This idea can be formalized by introducing a function $\phi_t: S \to S$ referred to either as the *flow* or the *evolution operator*.

The term 'evolution operator' is usually used in applications where ϕ_t describes the time development of the state of a real physical system. The word 'flow' is more frequently used when discussing the dynamics as a whole rather than the evolution of a particular point.

The function ϕ_t maps any $x_0 \in S$ onto the point $\phi_t(x_0)$ obtained by evolving for time t along a solution curve of (1.53) through x_0. Clearly, both existence and uniqueness of solutions are required for ϕ_t to be well defined. The point $\phi_t(x_0)$ is equal to $x(t + t_0)$ for any solution, $x(t)$, of (1.53) which satisfies $x(t_0) = x_0$ for some t_0. This property is illustrated in Fig. 1.42. As can be seen, it arises because the solutions of autonomous equations are related by translations in t. Thus the solution to $\dot{x} = X(x)$ with $x(t_0) = x_0$ is

$$x(t) = \phi_{t - t_0}(x_0). \tag{1.54}$$

A simple example of (1.54) is provided by the linear equation $\dot{x} = ax$. In this case

$$x(t) = \exp(a(t - t_0))x_0 \tag{1.55}$$

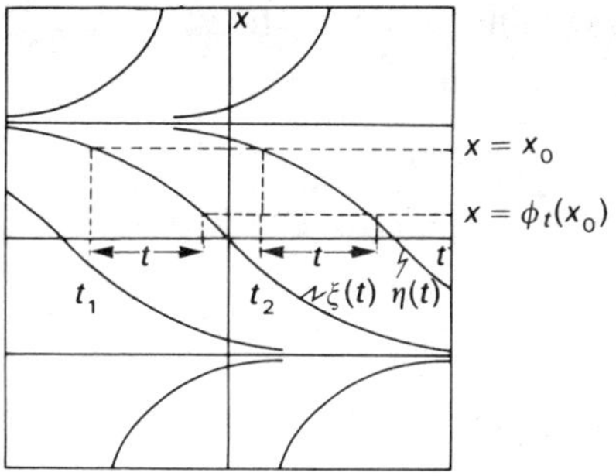

Fig. 1.42. Solution curves for $\dot{x} = \frac{1}{2}(x^2 - 1)$. Observe that the solutions $\xi(t)$ and $\eta(t)$ satisfy $\xi(t_1) = \eta(t_2) = x_0$, and $\xi(t + t_1) = \eta(t + t_2) = \phi_t(x_0)$, $t \in \mathrm{R}$.

so that ϕ_{t-t_0} is simply multiplication by $\exp(a(t - t_0))$. However, it is only for linear equations that the evolution operator takes this simple form. In fact, knowing ϕ_t is equivalent to having solved (1.53) and so finding it is of comparable difficulty.

The flow ϕ_t has simple properties which follow directly from its definition. Uniqueness conditions ensure that

$$\phi_{s+t}(x) = \phi_s(\phi_t(x)); \qquad s, t \in \mathrm{R} \tag{1.56}$$

providing both sides exist. In particular,

$$\phi_t(\phi_{-t}(x)) = \phi_{-t}(\phi_t(x)) = \phi_0(x) = x \tag{1.57}$$

and so

$$\phi_t^{-1} = \phi_{-t}. \tag{1.58}$$

For the flow in (1.55) the equalities (1.56) and (1.58) follow immediately from the properties of the exponential function. However, these relations are not always so apparent.

Example 1.5.1. Find the evolution operator ϕ_t for the equation

$$\dot{x} = x - x^2. \tag{1.59}$$

Verify (1.56) for this example.

Solution. The solutions to (1.59) satisfy

$$\int^{x} \frac{du}{u-u^2} = \ln\left|\frac{x}{x-1}\right| = t + C \tag{1.60}$$

for $x \neq 0$ or 1. This relation can be rearranged to give

$$\frac{x}{x-1} = K\mathrm{e}^{t}$$

with $K = \pm \mathrm{e}^{C}$, and thus

$$x(t) = K\mathrm{e}^{t}/(K\mathrm{e}^{t} - 1). \tag{1.61}$$

If we let $x = x_0$ at $t = 0$, then (1.61) implies $K = x_0/(x_0 - 1)$ and so

$$x(t) = \phi_t(x_0) = x_0\mathrm{e}^{t}/(x_0\mathrm{e}^{t} - x_0 + 1) \tag{1.62}$$

for t real and $x_0 \neq 0$ or 1. The points $x = 0$ and 1 were excluded in (1.60) because the integral is not defined for intervals including them. However, we see from (1.59) that they are the fixed points of the equation, which means that

$$\phi_t(0) = 0 \quad \text{and} \quad \phi_t(1) = 1, \tag{1.63}$$

for all $t \in \mathsf{R}$. The form for ϕ_t is given in (1.62) has precisely these properties and we can take

$$\phi_t(x) = x\mathrm{e}^{t}/(x\mathrm{e}^{t} - x + 1) \tag{1.64}$$

for *all* real x and t.

To check the basic property (1.56) of the evolution operator, observe

$$\phi_s(\phi_t(x)) = \phi_s(x_1) \tag{1.65}$$

where $x_1 = \phi_t(x)$. Thus

$$\phi_s(\phi_t(x)) = x_1\mathrm{e}^{s}/(x_1\mathrm{e}^{s} - x_1 + 1) \tag{1.66}$$

with

$$x_1 = x\mathrm{e}^{t}/(x\mathrm{e}^{t} - x + 1). \tag{1.67}$$

Substituting (1.67) into (1.66) gives

$$\phi_s(\phi_t(x)) = x\mathrm{e}^{s+t}/(x\mathrm{e}^{s+t} - x + 1) = \phi_{s+t}(x). \tag{1.68}$$ □

In Section 1.1, we saw that solutions of $\dot{x} = X(x)$ may not be defined for all real t. The following example illustrates how this is reflected in the domain of definition of $\phi_t(x)$.

Example 1.5.2. Find the evolution operator ϕ_t for the equation

$$\dot{x} = x^2 \tag{1.69}$$

and give the intervals of t on which it is defined for each real x.

Solution. Solutions to (1.69) satisfy

$$\int^x u^{-2}du = -x^{-1} = t + C \tag{1.70}$$

where C is a constant, in any interval which does not contain $x = 0$. If $x = x_0$ when $t = 0$, then $C = -x_0^{-1}$ and we obtain

$$x(t) = x_0/(1 - x_0 t), \qquad t \neq x_0^{-1}. \tag{1.71}$$

In terms of the evolution operator, (1.71) means that

$$\phi_t(x) = x/(1 - xt) \tag{1.72}$$

for any non-zero x. As in Example 1.5.1, $\phi_t(x)$ given in (1.72) is also valid at $x = 0$; i.e.

$$\phi_t(0) = 0 \quad \text{for all real } t, \tag{1.73}$$

as required by the fixed point at $x = 0$ in (1.69). Thus (1.72) is valid for all real x. However, $\phi_t(x)$ is not defined for all t; consider, for example, $t = x^{-1}$ in (1.72). In fact, the interval in t for which $\phi_t(x)$ is defined is determined by x as follows:

(a) $x > 0$: the entire evolution of x is given by (1.72) for $-\infty < t < x^{-1}$;
(b) $x = 0$: (1.73) shows that $\phi_t(0)$ is defined for $-\infty < t < \infty$;
(c) $x < 0$: (1.72) again describes the evolution of x but with $x^{-1} < t < \infty$.

In case (a), $\phi_t(x)$ increases from arbitrarily small, positive values at large negative t; through x at $t = 0$ and tends to infinity as $t \to x^{-1}$. Similarly, in case (c) $\phi_t(x)$ takes arbitrarily large, negative values for t close to x^{-1}; as t increases, x increases strictly and approaches zero as $t \to \infty$ (cf. Fig. 1.15). □

The evolution operator plays an analogous role in the plane. The autonomous nature of (1.19) again ensures that solutions are related by translations in t and ϕ_t maps $\mathbf{x} \in \mathsf{R}^2$ to the point obtained by evolving for time t from $\mathbf{x}$ according to $\dot{\mathbf{x}} = \mathbf{X}(\mathbf{x})$, i.e. ϕ_t: $\mathsf{R}^2 \to \mathsf{R}^2$. In these terms the trajectory passing through $\mathbf{x}$ is simply $\{\phi_t(\mathbf{x}) : t \text{ real}\}$ oriented by increasing t. As we shall see in Chapters 3 and 5, this notation can be a more useful description of the solutions.

In Section 2.5 we discuss a general method for obtaining the evolution operator for linear systems in the plane as a 2×2 matrix. For example, the solutions

$$x_1(t) = C_1 \mathrm{e}^{-t}, \qquad x_2(t) = C_2 \mathrm{e}^{-2t} \tag{1.74}$$

for $\dot{x}_1 = -x_1$, $\dot{x}_2 = -2x_2$ can be written as

$$\begin{bmatrix} x_1(t) \\ x_2(t) \end{bmatrix} = \begin{bmatrix} \mathrm{e}^{-t} & 0 \\ 0 & \mathrm{e}^{-2t} \end{bmatrix} \begin{bmatrix} C_1 \\ C_2 \end{bmatrix}. \tag{1.75}$$

It follows that

$$\phi_t = \begin{bmatrix} \mathrm{e}^{-t} & 0 \\ 0 & \mathrm{e}^{-2t} \end{bmatrix} \tag{1.76}$$

is the evolution operator for (1.74).

Exercises

Section 1.1

1. Show that a differential equation of the form

$$\dot{x} + p(t)x = q(t)$$

can be written as

$$\frac{\mathrm{d}}{\mathrm{d}t}(x\mathrm{e}^{P(t)}) = q(t)\mathrm{e}^{P(t)},$$

where $P(t) = \int^t p(s)\,\mathrm{d}s$. The function $\mathrm{e}^{P(t)}$ is called the integrating factor for the equation.

Use this observation to solve the following equations:
(a) $\dot{x} = x - t$; (b) $\dot{x} = x + \mathrm{e}^t$;
(c) $\dot{x} = -x \cot t + 5\mathrm{e}^{\cos t}$, $t \neq n\pi$, n integer.

2. A differential equation of the form

$$\dot{x} = f(x)g(t)$$

is said to be *separable*, because the solution passing through the point (t_0, x_0) of the t, x plane satisfies

$$\int_{x_0}^{x} \frac{du}{f(u)} = \int_{t_0}^{t} g(s)\,ds,$$

provided these integrals exist. The variables x and t are separated in this relation. Use this result to find solutions to:

(a) $\dot{x} = xt$; (b) $\dot{x} = -x/t, \quad t \neq 0$;
(c) $\dot{x} = -t/x, \quad x \neq 0$; (d) $\dot{x} = -x/\tanh t, \quad t \neq 0$.

3. A differential equation of the form

$$M(t, x) + N(t, x)\dot{x} = 0$$

is said to be *exact* if there is a function $F(t, x)$ with continuous second partial derivatives such that $\partial F/\partial t \equiv M(t, x)$ and $\partial F/\partial x \equiv N(t, x)$. Show that a necessary condition for such a function to exist is that

$$\frac{\partial M}{\partial x} = \frac{\partial N}{\partial t}$$

and that any solution to the differential equation satisfies

$$F(t, x) = \text{constant}.$$

Show that

$$\frac{x}{t} + [\ln(xt) + 1]\dot{x} = 0, \qquad t, x > 0$$

is an exact differential equation. Find $F(t, x)$ and plot several solution curves.

4. Use the calculus to find solutions for the following differential equations:

(a) $\dot{x} = x^2$; (b) $\dot{x} = \frac{1}{2}x^3$;
(c) $\dot{x} = \frac{1}{2}(x^2 - 1)$; (d) $\dot{x} = 3x^{2/3}$;
(e) $\dot{x} = \sqrt{(1 - x^2)}, \ |x| \leqslant 1$; (f) $\dot{x} = 2x^{1/2}, \ x \geqslant 0$.

Equations (e) and (f) are defined on the *closed* domains $D = \{(t, x) \,|\, |x| \leqslant 1\}$ and $D = \{(t, x) \,|\, x \geqslant 0\}$, respectively. Proposition 1.1.1 ensures existence of solutions on the *open* domains $D' = \{(t, x) \,|\, |x| < 1\}$ and $D' = \{t, x) \,|\, x > 0\}$ for these examples. Do the solutions you have found exist on the boundary of D?

5. Show how to construct infinitely many solutions, satisfying the initial condition $x(0) = 0$, for the differential equations given in Exercise 1.4 (d) and (f). Can the same be done for the equation in Exercise 1.4 (e) subject to the condition (a) $x(0) = 1$, (b) $x(0) = -1$? Explain your answer.

6. Suppose the differential equation $\dot{x} = X(t, x)$ has the property $X(t, x) = X(-t, -x)$. Prove that if $x = \xi(t)$ is a solution then so is $x = -\xi(-t)$. Find similar results on the symmetry of solutions when: (a) $X(t, x) = -X(-t, x)$; (b) $X(t, x) = -X(t, -x)$. Which of these symmetries appear in Figs. 1.1–1.8?

7. A differential equation of the form

$$\dot{x} = h(t, x) \tag{1}$$

is said to be *homogeneous* if $h(t, x)$ satisfies $h(\alpha t, \alpha x) \equiv h(t, x)$ for all non-zero real α. Show that the isoclines of such an equation are always straight lines through the origin of the t, x-plane.

Use this result to sketch the solution curves of

$$\dot{x} = e^{x/t}, \qquad t \neq 0. \tag{2}$$

Show that the change of variable $x = ut$ allows (1) to be written as a separable equation (Exercise 1.2) for u when $t \neq 0$. Does this result help to obtain the family of solution curves for (2)?

8. Sketch the family of solutions of the differential equation

$$\dot{x} = ax - bx^2, \qquad x > 0, \qquad a \text{ and } b > 0.$$

Obtain the sketch directly from the differential equation itself. Prove that $\dot{x}$ is increasing for $0 < x < a/2b$ and decreasing for $a/2b < x < a/b$. How does this result influence your sketches? How does $\dot{x}$ behave for $a/b < x < \infty$?

9. Show that the substitution $y = x^{-1}$, $x \neq 0$ allows

$$\dot{x} = ax - bx^2$$

to be written as a differential equation for y with the form described in Exercise 1.1. Solve this equation and show that

$$x(t) = ax_0/\{bx_0 + (a - bx_0)\exp(-a(t - t_0))\},$$

where $x(t_0) = x_0$. Verify that the sketches obtained in Exercise 1.8 agree with this result. Can you identify any new qualitative features of the solutions which are not apparent from the original differential equation?

10. Sketch the solution curves of the differential equations:
(a) $\dot{x} = x^2 - t^2 - 1$; (b) $\dot{x} = t - t/x$, $x \neq 0$;
(c) $\dot{x} = (2t + x)/(t - 2x)$, $t \neq 2x$; (d) $\dot{x} = x^2 + t^2$;
by using isoclines and the regions of convexity and concavity.

11. Obtain isoclines and sketch the family of solutions for the following differential equations without finding x as a function of t.
(a) $\dot{x} = x + t$; (b) $\dot{x} = x^3 - x$;
(c) $\dot{x} = xt^2$; (d) $\dot{x} = x \ln x$, $x > 0$;
(e) $\dot{x} = \sinh x$; (f) $\dot{x} = t(x + 1)/(t^2 + 1)$.
What geometrical feature do the isoclines of (b), (d) and (e) have in common? Finally, verify your results using calculus.

Section 1.2

12. Find the fixed points of the following autonomous differential equations:
(a) $\dot{x} = x + 1$; (b) $\dot{x} = x - x^3$; (c) $\dot{x} = \sinh(x^2)$;
(d) $\dot{x} = x^4 - x^3 - 2x^2$; (e) $\dot{x} = x^2 + 1$.
Determine the nature (attractor, repellor or shunt) of each fixed point and hence construct the phase portrait of each equation.

13. Which differential equations, in the following list, have the same phase portrait?
(a) $\dot{x} = \sinh x$; (b) $\dot{x} = ax$, $a > 0$; (c) $\dot{x} = \begin{cases} x \ln|x|, & x \neq 0 \\ 0, & x = 0; \end{cases}$
(d) $\dot{x} = \sin x$; (e) $\dot{x} = x^3 - x$; (f) $\dot{x} = \tanh x$.
Explain, in your own words, the significance of two differential equations having the same phase portrait.

14. Consider the parameter dependent differential equation

$$\dot{x} = (x - \lambda)(x^2 - \lambda), \quad \lambda \text{ real.}$$

Find all possible phase portraits that could occur for this equation together with the intervals of λ in which they occur.

15. How many distinct qualitative types of phase portrait can occur on the phase line for a differential equation with three fixed points? What is the formula for the number of distinct phase portraits in the general case with n fixed points?

16. Show that the phase portrait of

$$\dot{x} = (a - x)(b - x)$$

is qualitatively the same as that of

$$\dot{y} = y(y - c)$$

for all real a, b, c; $a \neq b, c \neq 0$. Show, however, that a transformation, $y = kx + l$, which takes the first equation into the second, exists if and only if $c = b - a$ or $a - b$.

17. Consider the differential equation

$$\dot{x} = x^3 + ax - b.$$

Show that there is a curve C in the a, b-plane which separates this plane into two regions A and B such that: if $(a, b) \in A$ the phase portrait consists of a single repellor and if $(a, b) \in B$ it has two repellors separated by an attractor. Let $a < 0$ be fixed; describe the change in configuration of the fixed points as b varies from $-\infty$ to ∞.

18. A substance γ is formed in a chemical reaction between substances α and β. In the reaction each gram of γ is produced by the combination of p grams of α and $q = 1 - p$ grams of β. The rate of formation of γ at any instant of time, t, is equal to the product of the masses of α and β that remain uncombined at that instant. If a grams of α and b grams of β are brought together at $t = 0$, show that the differential equation governing the mass, $x(t)$, of γ present at time $t > 0$ is

$$\dot{x} = (a - px)(b - qx).$$

Assume $a/p > b/q$ and construct the phase portrait for this equation. What is the maximum amount of γ that can possibly be produced in this experiment?

Section 1.3

19. Find the fixed points of the following systems of differential equations in the plane:

(a) $\dot{x}_1 = x_1(a - bx_2)$
$\dot{x}_2 = -x_2(c - dx_1)$
$a, b, c, d > 0$;

(b) $\dot{x}_1 = x_2$
$\dot{x}_2 = -\sin x_1$;

(c) $\dot{x}_1 = x_2$
$\dot{x}_2 = x_2(1 - x_1^2) - x_1$;

(d) $\dot{x}_1 = x_1(2 - x_1 - 2x_2)$
$\dot{x}_2 = x_2(2 - 2x_1 - x_2)$;

(e) $\dot{x}_1 = \sin x_1$
$\dot{x}_2 = \cos x_2$.

20. Sketch the following parametrized families of curves in the plane:
(a) $(x_1, x_2) = (a\cos t, a\sin t)$; (b) $(x_1, x_2) = (a\cos t, 2a\sin t)$;
(c) $(x_1, x_2) = (ae^t, be^{-2t})$; (d) $(x_1, x_2) = (ae^t + be^{-t}, ae^t - be^{-t})$;
(e) $(x_1, x_2) = (ae^t + be^{2t}, be^{2t})$;
where $a, b \in \mathbb{R}$. Find the systems of differential equations in the plane for which these curves form the phase portrait.

21. Use the sketches obtained in Exercise 1.20, to arrange the families of curves (a)–(e) in groups with the same type of fixed point at the origin of the x_1x_2-plane.

22. Consider the phase portrait of the system

$$\dot{x}_1 = X_1(x_1, x_2), \qquad \dot{x}_2 = X_2(x_1, x_2). \tag{1}$$

Show that the system

$$\dot{y}_1 = -X_1(y_1, y_2), \qquad \dot{y}_2 = -X_2(y_1, y_2)$$

has trajectories with the same shape but with the reverse orientation to those of (1). Verify your result by obtaining solutions for:

(a) $\dot{x}_1 = x_1$
$\dot{x}_2 = x_2$;

(b) $\dot{x}_1 = x_1$
$\dot{x}_2 = 2x_2$

and comparing with Figs. 1.23 and 1.24.

Section 1.4

23. Use the change of variable

$$x_1 = y_1 + y_2, \qquad x_2 = y_1 - y_2$$

to decouple the pair of differential equations

$$\dot{x}_1 = x_2, \qquad \dot{x}_2 = x_1.$$

Hence construct the phase portrait for the system.

24. Show that a solution of the system of differential equations

$$\dot{x}_1 = -2x_1, \qquad \dot{x}_2 = x_1 - 2x_2$$

satisfying $x_1(0) = 1$, $x_2(0) = 2$ is given by

$$x_1(t) = e^{-2t}, \qquad x_2(t) = e^{-2t}(t+2).$$

Prove that this solution is unique.

25. Consider the system of non-linear first-order differential equations

$$\dot{x}_1 = -x_2 + x_1(1 - x_1^2 - x_2^2), \qquad \dot{x}_2 = x_1 + x_2(1 - x_1^2 - x_2^2). \quad (1)$$

Use the change of variables $x_1 = r\cos\theta$, $x_2 = r\sin\theta$ to show that (1) is equivalent to the system

$$\dot{r} = r(1 - r^2), \qquad \dot{\theta} = 1. \quad (2)$$

Solve (2) with initial conditions $r(0) = r_0$ and $\theta(0) = \theta_0$ to obtain

$$r(t) = r_0/[r_0^2 + (1 - r_0^2)e^{-2t}]^{1/2}, \qquad \theta(t) = t + \theta_0.$$

26. (a) Suppose the autonomous system

$$\dot{\mathbf{x}} = \mathbf{X}(\mathbf{x}) \quad (1)$$

is invariant under the transformation $\mathbf{x} \to -\mathbf{x}$. Show that if $\boldsymbol{\xi}(t)$ satisfies (1) then so does $\boldsymbol{\eta}(t) = -\boldsymbol{\xi}(t)$.

(b) Suppose, instead, that $\mathbf{X}$ satisfies $\mathbf{X}(\mathbf{x}) = \mathbf{X}(-\mathbf{x})$. Show, in this case, that if $\boldsymbol{\xi}(t)$ is a solution to (1) then $\boldsymbol{\eta}(t) = -\boldsymbol{\xi}(-t)$ is also a solution.

Illustrate the relations obtained in (a) and (b) by examining typical trajectories $\boldsymbol{\xi}(t)$ and $\boldsymbol{\eta}(t)$ for:

(i) $\dot{x}_1 = x_1$, $\dot{x}_2 = x_1 + x_2$; (ii) $\dot{x}_1 = x_1^2$, $\dot{x}_2 = x_2^4$.

27. (a) Consider the non-autonomous equation

$$\dot{x} = X(t, x) \quad (1)$$

discussed in Section 1.1. Use the substitution $x_1 = t$, $x_2 = x$ to show that (1) is equivalent to the *autonomous* system

$$\dot{x}_1 = 1, \qquad \dot{x}_2 = X(x_1, x_2).$$

(b) Use the substitution $x_1 = x$, $x_2 = \dot{x}$ to show that any second-order equation $\ddot{x} = F(x, \dot{x})$ is equivalent to the autonomous system

$$\dot{x}_1 = x_2, \qquad \dot{x}_2 = F(x_1, x_2).$$

Use these observations to convert the following equations into equivalent first order *autonomous* systems:
(i) $\dot{x} = x - t$; (ii) $\dot{x} = xt$; (iii) $\ddot{x} + \sin x = 0$;
(iv) $\ddot{x} + 2a\dot{x} + bx = 0$; (v) $\ddot{x} + f(x)\dot{x} + g(x) = 0$.
Show that (b) is not a unique procedure by verifying that $\dot{x}_1 = x_2 - \int^{x_1} f(u)\,du$, $\dot{x}_2 = -g(x_1)$ gives alternative systems for (iv) and (v).

28. Reduce the following sets of equations to a system of *autonomous* first-order equations in an appropriate number of variables:
(a) $\ddot{x} + x = 1$, $\ddot{y} + \dot{y} + y = 0$; (b) $\dot{x} + t = x$, $\dot{y} + y^3 = t$;
(c) $\ddot{x} + tx + 1 = 0$, $\ddot{y} + t^2\dot{x}^2 + x + y = 0$.

29. Use the method of isoclines to sketch the phase portraits of the following systems:
(a) $\dot{x}_1 = x_1 + x_2$, $\dot{x}_2 = x_1^2$; (b) $\dot{x}_1 = x_2^3$, $\dot{x}_2 = x_1 - x_2^2$;
(c) $\dot{x}_1 = \ln x_1$; $\dot{x}_2 = x_2$, $x_1 > 0$.

30. The following systems of differential equations in the plane all have a single fixed point. Find the fixed point and use the method of isoclines to determine which systems do *not* have closed orbits in their phase portraits.
(a) $\dot{x}_1 = x_2 - 1$, $\dot{x}_2 = -(x_1 - 2)$; (b) $\dot{x}_1 = x_1 + x_2$, $\dot{x}_2 = x_1$; (c) $\dot{x}_1 = x_1 - x_2$, $\dot{x}_2 = x_1 + x_2$.
Confirm your results using calculus to find the solutions of $dx_2/dx_1 = \dot{x}_2/\dot{x}_1$.

31. Consider the differential equation

$$\ddot{x} + \dot{x}^2 + x = 0. \tag{1}$$

Use Exercise 1.27 (b) to convert it to first-order form. Show that the isocline of slope k

$$\mathscr{I}_k = \{(x_1, x_2) \mid \dot{x}_2 = k\dot{x}_1\}$$

is a parabola with vertex $(k^2/4, -k/2)$ and that these vertices themselves lie on a parabola $x_2^2 = x_1$. Use the substitution $x_2^2 = w$ to solve for x_2 as a function of x_1 and hence sketch the phase portrait of (1).

Section 1.5

32. Show that the evolution operator of the differential equation $\dot{x} = x - x^3$ for $x > 1$ is given by

$$\phi_t(x) = xe^t/\sqrt{(x^2e^{2t} - x^2 + 1)}.$$

Check that $\phi_s(\phi_t(x)) = \phi_{s+t}(x)$.

33. Find the evolution operators of the differential equations:
(a) $\dot{x} = \tanh x$; (b) $\dot{x} = x \ln x$, $x > 0$.

34. Find the evolution operators of the following systems:
(a) $\dot{x}_1 = x_1$, $\dot{x}_2 = x_1 - x_2$; (b) $\dot{x}_1 = x_1 x_2$, $\dot{x}_2 = x_2^2$.
Give the intervals of time for which each operator is defined and verify that $\phi_{t+s}(\mathbf{x}) = \phi_t(\phi_s(\mathbf{x}))$ provided t, s and $t+s$ belong to the same interval of definition.

35. Let ϕ_t describe a flow in the plane. Draw phase portraits with at most two fixed points that satisfy each of the constraints given below:
(a) $\lim_{t\to\infty} \phi_t(\mathbf{x}) = \mathbf{0}$ for all points $\mathbf{x}$ of the plane;
(b) there is a point $\mathbf{x}_0$ such that $\lim_{t\to\infty} \phi_t(\mathbf{x}_0) = \lim_{t\to-\infty} \phi_t(\mathbf{x}_0) = \mathbf{0}$;
(c) $\lim_{t\to-\infty} \phi_t(\mathbf{x}) = \mathbf{0}$ for all points $\mathbf{x}$ of the plane;
(d) there is a trajectory through $\mathbf{x}_0$ such that $\lim_{t\to\infty}\phi_t(\mathbf{x}_0) = \mathbf{0}$ and a trajectory through $\mathbf{x}_0'$ such that $\lim_{t\to-\infty} \phi_t(\mathbf{x}_0') = \mathbf{0}$.

36. Given the differential equation $\dot{x} = xt$, $x, t \geqslant 0$, prove that the evolution of the point x_0 at $t = t_0$ does not depend *only* on x_0 and $t - t_0$ as in the autonomous case by showing that

$$\phi(t, t_0, x_0) = x_0 e^{(t-t_0)(t+t_0)/2}.$$

CHAPTER TWO

Linear systems

A system $\dot{\mathbf{x}} = \mathbf{X}(\mathbf{x})$, where $\mathbf{x}$ is a vector in R^n, is called a *linear system* of dimension n, if $\mathbf{X}: \mathsf{R}^n \to \mathsf{R}^n$ is a linear mapping. We will show that only a finite number of qualitatively different phase portraits can arise for linear systems. To do this we will first consider how such a system is affected by a linear change of variables.

2.1 Linear changes of variable

If the mapping $\mathbf{X}: \mathsf{R}^n \to \mathsf{R}^n$, where $\mathsf{R}^n = \{(x_1, \ldots, x_n) | x_i \in \mathsf{R}, i = 1, \ldots, n\}$, is *linear*, then it can be written in the matrix form

$$X(x) = \begin{bmatrix} X_1(x_1, \ldots, x_n) \\ \vdots \\ X_n(x_1, \ldots, x_n) \end{bmatrix} = \begin{bmatrix} a_{11} & \ldots & a_{1n} \\ \vdots & & \vdots \\ a_{n1} & \ldots & a_{nn} \end{bmatrix} \begin{bmatrix} x_1 \\ \vdots \\ x_n \end{bmatrix}. \tag{2.1}$$

Correspondingly, $\dot{\mathbf{x}} = \mathbf{X}(\mathbf{x})$ becomes

$$\dot{x} = X(x) = Ax, \tag{2.2}$$

where A is the *coefficient matrix*. Each component $X_i (i = 1, \ldots, n)$ of $\dot{\mathbf{x}}$ is a linear function of the variables $x_1, \ldots, x_n$. These variables are, of course, simply the coordinates of $\mathbf{x} = (x_1, \ldots, x_n)$ relative to the natural basis in R^n (i.e. $\{\mathbf{e}_i\}_{i=1}^n$, where $\mathbf{e}_i = (0, \ldots, 0, 1, 0, \ldots, 0)$ with 1 in the ith position). Thus

$$\mathbf{x} = \sum_{i=1}^{n} x_i \mathbf{e}_i \quad \text{and} \quad \mathbf{X}(\mathbf{x}) = \sum_{i=1}^{n} X_i(\mathbf{x})\mathbf{e}_i. \tag{2.3}$$

In order to make a change of variables we must express each $x_i(i = 1, \ldots, n)$ as a function of the new variables. We consider the effect on (2.2), of making a linear change of variable,

$$x_i = \sum_{j=1}^{n} m_{ij} y_j \, (i = 1, \ldots, n) \quad \text{or} \quad \boldsymbol{x} = \boldsymbol{M}\boldsymbol{y} \tag{2.4}$$

where m_{ij} is a real constant for all i and j. Of course, there must be a unique set of new variables $(y_1, \ldots, y_n)$ corresponding to a given set of old ones $(x_1, \ldots, x_n)$, and vice versa. This means that (2.4) must be a bijection and therefore $\boldsymbol{M}$ is a non-singular matrix. It follows that the columns $\boldsymbol{m}_i$, $i = 1, \ldots, n$, of $\boldsymbol{M}$ are linearly independent. Equation (2.4) implies

$$\boldsymbol{x} = \sum_{i=1}^{n} y_i \boldsymbol{m}_i, \tag{2.5}$$

and we recognize $y_1, \ldots, y_n$ as the coordinates of $\mathbf{x} \in \mathrm{R}^n$ relative to the new basis $\{\mathbf{m}_i\}_{i=1}^{n}$.

It is easy to express (2.2) in terms of the new variables; we find

$$\dot{\boldsymbol{x}} = \boldsymbol{M}\dot{\boldsymbol{y}} = \boldsymbol{A}\boldsymbol{M}\boldsymbol{y} \tag{2.6}$$

so that

$$\dot{\boldsymbol{y}} = \boldsymbol{B}\boldsymbol{y}, \tag{2.7}$$

with

$$\boldsymbol{B} = \boldsymbol{M}^{-1}\boldsymbol{A}\boldsymbol{M}. \tag{2.8}$$

Thus the coefficient matrix $\boldsymbol{B}$ of the transformed system is *similar* to $\boldsymbol{A}$.

Example 2.1.1. The system

$$\dot{x}_1 = x_2, \qquad \dot{x}_2 = x_1 \tag{2.9}$$

is transformed into the decoupled system

$$\dot{y}_1 = y_1, \qquad \dot{y}_2 = -y_2 \tag{2.10}$$

by the change of variable

$$x_1 = y_1 + y_2; \qquad x_2 = y_1 - y_2. \tag{2.11}$$

Use this result to illustrate (2.8).

Solution. Expressing these systems of equations in matrix form we obtain

$$\begin{bmatrix}\dot{x}_1\\ \dot{x}_2\end{bmatrix}=\begin{bmatrix}0 & 1\\ 1 & 0\end{bmatrix}\begin{bmatrix}x_1\\ x_2\end{bmatrix},\qquad \begin{bmatrix}x_1\\ x_2\end{bmatrix}=\begin{bmatrix}1 & 1\\ 1 & -1\end{bmatrix}\begin{bmatrix}y_1\\ y_2\end{bmatrix}\quad\text{and}$$

$$\begin{bmatrix}\dot{y}_1\\ \dot{y}_2\end{bmatrix}=\begin{bmatrix}1 & 0\\ 0 & -1\end{bmatrix}\begin{bmatrix}y_1\\ y_2\end{bmatrix}.$$

Thus

$$\boldsymbol{A}=\begin{bmatrix}0 & 1\\ 1 & 0\end{bmatrix},\qquad \boldsymbol{M}=\begin{bmatrix}1 & 1\\ 1 & -1\end{bmatrix}\quad\text{and}\quad \boldsymbol{B}=\begin{bmatrix}1 & 0\\ 0 & -1\end{bmatrix}.$$

Observe that $\boldsymbol{AM}=\boldsymbol{MB}$ so that (2.8) is satisfied. □

Example 2.1.2. Find the matrix representation of the linear system

$$\dot{x}_1 = x_1 + 2x_2,\qquad \dot{x}_2 = 2x_2 \tag{2.12}$$

under the change of variables

$$x_1 = y_1 + 2y_2,\qquad x_2 = y_2. \tag{2.13}$$

What is the basis for which y_1, y_2 are the corresponding coordinates?

Solution. The change of variables (2.13) can be written in the form

$$\begin{bmatrix}x_1\\ x_2\end{bmatrix}=\begin{bmatrix}1 & 2\\ 0 & 1\end{bmatrix}\begin{bmatrix}y_1\\ y_2\end{bmatrix} \tag{2.14}$$

so that

$$\boldsymbol{M}=\begin{bmatrix}1 & 2\\ 0 & 1\end{bmatrix}. \tag{2.15}$$

The system (2.12) has the matrix form

$$\dot{\boldsymbol{x}}=\begin{bmatrix}1 & 2\\ 0 & 2\end{bmatrix}\boldsymbol{x}=\boldsymbol{Ax}. \tag{2.16}$$

Equations (2.7) and (2.8) give

$$\boldsymbol{B}=\boldsymbol{M}^{-1}\boldsymbol{AM}=\begin{bmatrix}1 & 0\\ 0 & 2\end{bmatrix}\quad\text{and}\quad \dot{\boldsymbol{y}}=\boldsymbol{By} \tag{2.17}$$

as the required matrix representation.

The basis with coordinates y_1, y_2 is given by the columns of $\boldsymbol{M}$ in (2.15), that is $\{(1, 0), (2, 1)\}$. □

Similarity is an equivalence relation on the set of $n \times n$ real matrices (see Exercise 2.2) and it follows that this set can be disjointly decomposed into equivalence or *similarity classes*. For any two matrices $\boldsymbol{A}$ and $\boldsymbol{B}$ in the same similarity class, the solutions of the systems $\dot{\boldsymbol{x}} = \boldsymbol{A}\boldsymbol{x}$ and $\dot{\boldsymbol{y}} = \boldsymbol{B}\boldsymbol{y}$ are related by $\boldsymbol{x} = \boldsymbol{M}\boldsymbol{y}$ if $\boldsymbol{M}^{-1}\boldsymbol{A}\boldsymbol{M} = \boldsymbol{B}$. Thus, if one such system can be solved, solutions can be obtained for each member of the class.

In Section 1.4.1, we saw that decoupled or partially decoupled systems could be solved easily and we are led to consider whether each similarity class contains at least one correspondingly simple (e.g. diagonal or triangular) member. The answer to this algebraic problem is known and we will illustrate it for $n = 2$ in the next section.

2.2 Similarity types for 2 × 2 real matrices

For each positive integer n, there are infinitely many similarity classes of $n \times n$ real matrices. These similarity classes can be grouped into just finitely many types. In the following proposition we give these types for $n = 2$.

Proposition 2.2.1. Let $\boldsymbol{A}$ be a real 2×2 matrix, then there is a real, non-singular matrix $\boldsymbol{M}$ such that $\boldsymbol{J} = \boldsymbol{M}^{-1}\boldsymbol{A}\boldsymbol{M}$ is one of the types:

$$\begin{aligned}&\text{(a)}\quad \begin{bmatrix}\lambda_1 & 0\\ 0 & \lambda_2\end{bmatrix},\quad \lambda_1 > \lambda_2; &&\text{(b)}\quad \begin{bmatrix}\lambda_0 & 0\\ 0 & \lambda_0\end{bmatrix};\\ &\text{(c)}\quad \begin{bmatrix}\lambda_0 & 1\\ 0 & \lambda_0\end{bmatrix}; &&\text{(d)}\quad \begin{bmatrix}\alpha & -\beta\\ \beta & \alpha\end{bmatrix},\quad \beta > 0\end{aligned} \tag{2.18}$$

where λ_0, λ_1, λ_2, α, β are real numbers.

The matrix $\boldsymbol{J}$ is said to be the *Jordan form* of $\boldsymbol{A}$. The eigenvalues of the matrix $\boldsymbol{A}$ (and $\boldsymbol{J}$) are the values of λ for which

$$p_A(\lambda) = \lambda^2 - \mathrm{tr}(\boldsymbol{A})\lambda + \det(\boldsymbol{A}) = 0. \tag{2.19}$$

Here $\mathrm{tr}(\boldsymbol{A}) = a_{11} + a_{22}$ is the trace of $\boldsymbol{A}$ and $\det(\boldsymbol{A}) = a_{11}a_{22} - a_{12}a_{21}$ is its determinant. Thus the eigenvalues of $\boldsymbol{A}$ are

$$\lambda_1 = \tfrac{1}{2}(\mathrm{tr}(\boldsymbol{A}) + \sqrt{\Delta}) \quad\text{and}\quad \lambda_2 = \tfrac{1}{2}(\mathrm{tr}(\boldsymbol{A}) - \sqrt{\Delta}) \tag{2.20}$$

with

$$\Delta = (\text{tr}(\boldsymbol{A}))^2 - 4\det(\boldsymbol{A}). \tag{2.21}$$

It is the nature of the eigenvalues: real distinct ($\Delta > 0$), real equal ($\Delta = 0$) and complex ($\Delta < 0$) that determine the type of the Jordan form $\boldsymbol{J}$ of $\boldsymbol{A}$.

(*a*) *Real distinct eigenvalues* ($\Delta > 0$)

The eigenvectors $\boldsymbol{u}_1, \boldsymbol{u}_2$ of $\boldsymbol{A}$ are given by

$$\boldsymbol{A}\boldsymbol{u}_i = \lambda_i \boldsymbol{u}_i, \qquad (i = 1, 2) \tag{2.22}$$

with $\lambda_1, \lambda_2, \lambda_1 > \lambda_2$, the distinct eigenvalues. Let

$$\boldsymbol{M} = [\boldsymbol{u}_1 \;\vdots\; \boldsymbol{u}_2] \tag{2.23}$$

be the matrix with the eigenvectors $\boldsymbol{u}_1$, $\boldsymbol{u}_2$ as columns. Then

$$\boldsymbol{A}\boldsymbol{M} = [\boldsymbol{A}\boldsymbol{u}_1 \;\vdots\; \boldsymbol{A}\boldsymbol{u}_2] = [\lambda_1 \boldsymbol{u}_1 \;\vdots\; \lambda_2 \boldsymbol{u}_2] = \boldsymbol{M}\boldsymbol{J} \tag{2.24}$$

where

$$\boldsymbol{J} = \begin{bmatrix} \lambda_1 & 0 \\ 0 & \lambda_2 \end{bmatrix}.$$

For distinct eigenvalues the eigenvectors $\boldsymbol{u}_1$ and $\boldsymbol{u}_2$ are linearly independent and therefore $\boldsymbol{M}$ is non-singular. Thus

$$\boldsymbol{M}^{-1}\boldsymbol{A}\boldsymbol{M} = \boldsymbol{J} = \begin{bmatrix} \lambda_1 & 0 \\ 0 & \lambda_2 \end{bmatrix}. \tag{2.25}$$

(*b*) *Equal eigenvalues* ($\Delta = 0$)

Equation (2.20) gives $\lambda_1 = \lambda_2 = \frac{1}{2}\text{tr}(\boldsymbol{A}) = \lambda_0$, and we must consider the following possibilities.

(*i*) ***A*** *is diagonal*

$$\boldsymbol{A} = \begin{bmatrix} \lambda_0 & 0 \\ 0 & \lambda_0 \end{bmatrix} = \lambda_0 \boldsymbol{I} \tag{2.26}$$

which is (2.18(b)). In this case for any non-singular matrix $\boldsymbol{M}$, $\boldsymbol{M}^{-1}\boldsymbol{A}\boldsymbol{M} = \boldsymbol{A}$. Therefore the matrix $\boldsymbol{A}$ is only similar to itself and hence is in a similarity class of its own.

(*ii*) ***A*** *is not diagonal*

In this case, since $\lambda_1 = \lambda_2 = \lambda_0$, $\text{rank}(\boldsymbol{A} - \lambda_0 \boldsymbol{I}) = 1$ and there are *not* two linearly independent eigenvectors. Let $\boldsymbol{u}_0$ be an eigenvector of $\boldsymbol{A}$.

If we take $\boldsymbol{m}_1 = \boldsymbol{u}_0$ and choose $\boldsymbol{m}_2$ so that $\boldsymbol{M} = [\boldsymbol{u}_0 \mid \boldsymbol{m}_2]$ is a non-singular matrix,

$$\boldsymbol{AM} = [\lambda_0 \boldsymbol{u}_0 \mid \boldsymbol{Am}_2] = \boldsymbol{M}[\lambda_0 \boldsymbol{e}_1 \mid \boldsymbol{M}^{-1}\boldsymbol{Am}_2], \tag{2.27}$$

where $\boldsymbol{e}_1$ is the first column of $\boldsymbol{I}$.

The matrices $\boldsymbol{A}$ and $\boldsymbol{M}^{-1}\boldsymbol{AM}$ have the same eigenvalues, and so

$$\boldsymbol{M}^{-1}\boldsymbol{AM} = \begin{bmatrix} \lambda_0 & C \\ 0 & \lambda_0 \end{bmatrix} \tag{2.28}$$

for some non-zero C. However, the simple modification of $\boldsymbol{M}$ to

$$\boldsymbol{M}_1 = \boldsymbol{M}\begin{bmatrix} 1 & 0 \\ 0 & C^{-1} \end{bmatrix} \tag{2.29}$$

results in

$$\boldsymbol{M}_1^{-1}\boldsymbol{AM}_1 = \begin{bmatrix} \lambda_0 & 1 \\ 0 & \lambda_0 \end{bmatrix}$$

which is (2.18(c)).

(c) *Complex eigenvalues* ($\Delta < 0$)

We can write $\lambda_1 = \alpha + i\beta$ and $\lambda_2 = \alpha - i\beta$, where $\alpha = \frac{1}{2}\operatorname{tr}(\boldsymbol{A})$ and $\beta = +\frac{1}{2}\sqrt{(-\Delta)}$. We wish to show that there is a non-singular matrix $\boldsymbol{M} = [\boldsymbol{m}_1 \mid \boldsymbol{m}_2]$, such that $\boldsymbol{M}^{-1}\boldsymbol{AM}$ is given by (2.18(d)), or equivalently that

$$\boldsymbol{AM} = \boldsymbol{M}\begin{bmatrix} \alpha & -\beta \\ \beta & \alpha \end{bmatrix}. \tag{2.30}$$

Partitioning $\boldsymbol{M}$ into its columns, we obtain

$$[\boldsymbol{Am}_1 \mid \boldsymbol{Am}_2] = [\alpha \boldsymbol{m}_1 + \beta \boldsymbol{m}_2 \mid -\beta \boldsymbol{m}_1 + \alpha \boldsymbol{m}_2]$$

or

$$[(\boldsymbol{A} - \alpha \boldsymbol{I})\boldsymbol{m}_1 - \beta \boldsymbol{I}\boldsymbol{m}_2 \mid \beta \boldsymbol{I}\boldsymbol{m}_1 + (\boldsymbol{A} - \alpha \boldsymbol{I})\boldsymbol{m}_2] = [\boldsymbol{0} \mid \boldsymbol{0}]. \tag{2.31}$$

This matrix equation can be written as four homogeneous linear equations for the unknown elements of $\boldsymbol{M}$, i.e.

$$\left[\begin{array}{c|c} \boldsymbol{A} - \alpha \boldsymbol{I} & -\beta \boldsymbol{I} \\ \hline \beta \boldsymbol{I} & \boldsymbol{A} - \alpha \boldsymbol{I} \end{array}\right] \left[\begin{array}{c} \boldsymbol{m}_1 \\ \hline \boldsymbol{m}_2 \end{array}\right] = \left[\begin{array}{c} \boldsymbol{0} \\ \hline \boldsymbol{0} \end{array}\right]. \tag{2.32}$$

To solve these equations, let $\boldsymbol{P}$ be the coefficient matrix in (2.32) and let

$$\boldsymbol{Q} = \left[\begin{array}{c|c} \boldsymbol{A}-\alpha\boldsymbol{I} & \beta\boldsymbol{I} \\ \hline -\beta\boldsymbol{I} & \boldsymbol{A}-\alpha\boldsymbol{I} \end{array}\right]. \tag{2.33}$$

Now observe that

$$\boldsymbol{PQ} = \left[\begin{array}{c|c} p_A(\boldsymbol{A}) & \boldsymbol{0} \\ \hline \boldsymbol{0} & p_A(\boldsymbol{A}) \end{array}\right] \tag{2.34}$$

where $p_A(\lambda) = \lambda^2 - 2\alpha\lambda + (\alpha^2 + \beta^2)$ is the characteristic polynomial of $\boldsymbol{A}$. The Cayley–Hamilton theorem (Hartley and Hawkes, 1970) states that $p_A(\boldsymbol{A}) = \boldsymbol{0}$ (see Exercise 2.7) and we conclude that

$$\boldsymbol{PQ} = \boldsymbol{0}, \tag{2.35}$$

the 4×4 null matrix. Thus, the columns of $\boldsymbol{Q}$ must be solutions to (2.32). The first column of $\boldsymbol{Q}$ gives

$$\boldsymbol{M} = \begin{bmatrix} a_{11}-\alpha & -\beta \\ a_{21} & 0 \end{bmatrix}. \tag{2.36}$$

Note that the discriminant Δ of $p_A(\lambda) = 0$ can be written

$$\Delta = (a_{11} - a_{22})^2 + 4a_{12}a_{21}. \tag{2.37}$$

If $\Delta < 0$ then $a_{12}a_{21} \neq 0$ and hence $a_{21} \neq 0$. Further, $\beta = +\frac{1}{2}\sqrt{(-\Delta)} \neq 0$ and we conclude that $\det(\boldsymbol{M}) = \beta a_{21} \neq 0$. Thus, (2.36) provides a non-singular matrix $\boldsymbol{M}$ such that $\boldsymbol{M}^{-1}\boldsymbol{AM}$ is given by (2.18(d)).

Any given 2×2 real matrix $\boldsymbol{A}$ falls into one, and only one, of the cases set out above (see Exercise 2.6) and Proposition 2.2.1 is justified. Let us consider some simple illustrations of this result.

Example 2.2.1. Find the Jordan forms of each of the following matrices:

$$\text{(a)}\ \begin{bmatrix} 1 & 2 \\ 1 & 1 \end{bmatrix}; \quad \text{(b)}\ \begin{bmatrix} 2 & 1 \\ -2 & 4 \end{bmatrix}; \quad \text{(c)}\ \begin{bmatrix} 3 & -1 \\ 1 & 1 \end{bmatrix}.$$

Solution. The pairs of eigenvalues of the matrices (a)–(c) are $(\lambda_1, \lambda_2) = (1 + \sqrt{2}, 1 - \sqrt{2})$, $(3 + i, 3 - i)$ and $(2, 2)$ respectively. Thus, the

Jordan forms are

$$
\begin{aligned}
&\text{(a)} \quad \begin{bmatrix} \lambda_1 & 0 \\ 0 & \lambda_2 \end{bmatrix} = \begin{bmatrix} 1+\sqrt{2} & 0 \\ 0 & 1-\sqrt{2} \end{bmatrix} \\
&\text{(b)} \quad \begin{bmatrix} \alpha & -\beta \\ \beta & \alpha \end{bmatrix} = \begin{bmatrix} 3 & -1 \\ 1 & 3 \end{bmatrix} \qquad (2.38) \\
&\text{(c)} \quad \begin{bmatrix} \lambda_0 & 1 \\ 0 & \lambda_0 \end{bmatrix} = \begin{bmatrix} 2 & 1 \\ 0 & 2 \end{bmatrix}
\end{aligned}
$$

respectively. Notice (2.18 (c)) is chosen for (2.38 (c)) above because the original matrix is non-diagonal. □

Example 2.2.2. Find a matrix M which converts each of the matrices in Example 2.2.1 into their appropriate Jordan forms.

Solution. Matrix (a) has real distinct eigenvalues and so the columns of M can be taken as the two corresponding eigenvectors. The eigenvector $u_1 = \begin{bmatrix} u_{11} \\ u_{21} \end{bmatrix}$ satisfies $[A-(1+\sqrt{2})I]u_1 = \mathbf{0}$ which implies only that $u_{11}\sqrt{2} = 2u_{21}$. We can take $u_1 = \begin{bmatrix} \sqrt{2} \\ 1 \end{bmatrix}$. Similarly, u_2 corresponding to $\lambda_2 = 1-\sqrt{2}$ can be taken as $\begin{bmatrix} -\sqrt{2} \\ 1 \end{bmatrix}$. Thus

$$M = \begin{bmatrix} \sqrt{2} & -\sqrt{2} \\ 1 & 1 \end{bmatrix} \quad \text{and} \quad M^{-1}\begin{bmatrix} 1 & 2 \\ 1 & 1 \end{bmatrix}M = \begin{bmatrix} 1+\sqrt{2} & 0 \\ 0 & 1-\sqrt{2} \end{bmatrix}.$$

Matrix (b) has complex conjugate eigenvalues $3+i, 3-i$ and Equation (2.36) gives an explicit form for M. In this case, we find

$$M = \begin{bmatrix} -1 & -1 \\ -2 & 0 \end{bmatrix}$$

and verify that

$$\begin{bmatrix} 2 & 1 \\ -2 & 4 \end{bmatrix}M = M\begin{bmatrix} 3 & -1 \\ 1 & 3 \end{bmatrix}.$$

Finally, matrix (c) has equal eigenvalues and a single eigenvector

$\begin{bmatrix} 1 \\ 1 \end{bmatrix}$ obtained by solving

$$\left\{\begin{bmatrix} 3 & -1 \\ 1 & 1 \end{bmatrix} - 2\begin{bmatrix} 1 & 0 \\ 0 & 1 \end{bmatrix}\right\}\begin{bmatrix} u_{10} \\ u_{20} \end{bmatrix} = \begin{bmatrix} 0 \\ 0 \end{bmatrix}.$$

This gives the first column of $\boldsymbol{M}$. Provided the second column is chosen to make $\boldsymbol{M}$ non-singular, matrix (c) can be reduced to upper triangular form. For example, let

$$\boldsymbol{M} = \begin{bmatrix} 1 & -1 \\ 1 & 0 \end{bmatrix};$$

this is a simple choice which makes $\det(\boldsymbol{M}) = 1$, then

$$\boldsymbol{M}^{-1}\begin{bmatrix} 3 & -1 \\ 1 & 1 \end{bmatrix}\boldsymbol{M} = \begin{bmatrix} 2 & -1 \\ 0 & 2 \end{bmatrix}.$$

This is *not* the Jordan form given in (2.38), however, if we replace $\boldsymbol{M}$ by $\boldsymbol{M}_1$ as suggested in (2.29), i.e.

$$\boldsymbol{M}_1 = \boldsymbol{M}\begin{bmatrix} 1 & 0 \\ 0 & -1 \end{bmatrix} = \begin{bmatrix} 1 & 1 \\ 1 & 0 \end{bmatrix},$$

The result $\boldsymbol{J} = \begin{bmatrix} 2 & 1 \\ 0 & 2 \end{bmatrix}$ is achieved. □

We will discuss the Jordan forms of matrices with dimension greater than two in Section 2.7, but first let us examine the significance of the above results for linear systems in the plane. We can summarize the situation as follows:

(a) every two dimensional linear system

$$\dot{\boldsymbol{x}} = \boldsymbol{A}\boldsymbol{x} \tag{2.39}$$

can be transformed into an equivalent *canonical system*

$$\dot{\boldsymbol{y}} = \boldsymbol{J}\boldsymbol{y}, \tag{2.40}$$

where $\boldsymbol{J} = \boldsymbol{M}^{-1}\boldsymbol{A}\boldsymbol{M}$ is the Jordan form of $\boldsymbol{A}$ and

$$\boldsymbol{x} = \boldsymbol{M}\boldsymbol{y}; \tag{2.41}$$

(b) the Jordan matrix $\boldsymbol{J}$ must belong to one of the four types given in the Proposition 2.2.1.

We can now find solutions to *any* system like (2.39) by solving the corresponding canonical system (2.40) for $\boldsymbol{y}$ and using (2.41) to find $\boldsymbol{x}$.

2.3 Phase portraits for canonical systems in the plane

A linear system $\dot{\boldsymbol{x}} = \boldsymbol{A}\boldsymbol{x}$ is said to be *simple* if the matrix $\boldsymbol{A}$ is non-singular, (i.e. $\det \boldsymbol{A} \neq 0$ and $\boldsymbol{A}$ has non-zero eigenvalues). The only solution to

$$\boldsymbol{A}\boldsymbol{x} = \boldsymbol{0} \tag{2.42}$$

is then $\boldsymbol{x} = \boldsymbol{0}$ and the system has a single isolated fixed point at the origin of the phase plane. The canonical system corresponding to a simple linear system is also simple because the eigenvalues of $\boldsymbol{A}$ and $\boldsymbol{J}$ are the same.

2.3.1 *Simple canonical systems*

(*a*) *Real, distinct eigenvalues*

In this case $\boldsymbol{J}$ is given by (2.18(a)) with λ_1 and λ_2 non-zero. Thus

$$\dot{y}_1 = \lambda_1 y_1, \qquad \dot{y}_2 = \lambda_2 y_2 \tag{2.43}$$

and hence

$$y_1(t) = C_1 e^{\lambda_1 t}, \qquad y_2(t) = C_2 e^{\lambda_2 t}, \tag{2.44}$$

where C_1, C_2 are real constants.

If λ_1 and λ_2 have the *same* sign phase portraits like those in Fig. 2.1 arise. The fixed point at the origin of the $y_1 y_2$-plane is said to be a

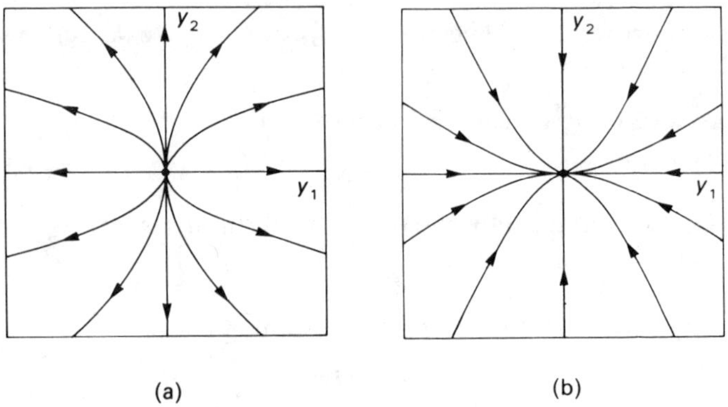

Fig. 2.1. Real distinct eigenvalues of the same sign give rise to nodes: (a) unstable ($\lambda_1 > \lambda_2 > 0$); (b) stable ($\lambda_2 < \lambda_1 < 0$).

node. When all the trajectories are oriented towards (away from) the origin the node is said to be *stable* (*unstable*). The shape of the trajectories is determined by the ratio $\gamma = \lambda_2/\lambda_1$. Notice that (2.43) and (2.44) give

$$\frac{dy_2}{dy_1} = Ky_1^{(\gamma-1)}, \tag{2.45}$$

where $K = \gamma C_2/C_1^{\gamma}$. Therefore, as $y_1 \to 0$

$$\frac{dy_2}{dy_1} \to \begin{cases} 0, & \text{if } \gamma > 1 \\ \infty, & \text{if } \gamma < 1 \end{cases}. \tag{2.46}$$

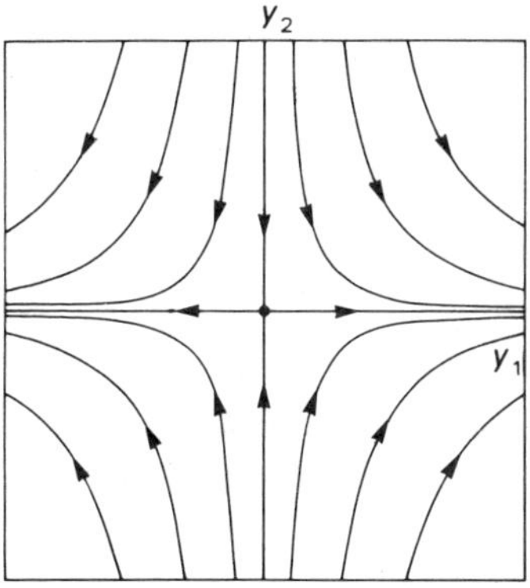

Fig. 2.2. Real eigenvalues of opposite sign ($\lambda_2 < 0 < \lambda_1$) give rise to saddle points.

If λ_1 and λ_2 have *opposite* signs, the phase portrait in Fig. 2.2 occurs. The coordinate axes (excluding the origin) are the unions of special trajectories, called *separatrices*. These are the only trajectories that are radial straight lines. A particular coordinate axis contains a pair of separatrices (remember the origin is a trajectory in its own right) which are directed towards (away from) the origin if the corresponding eigenvalue is negative (positive). The remaining trajectories have the separatrices as asymptotes; first approaching the fixed point as t increases from $-\infty$, passing through a point of closest approach and finally moving away again. In this case the origin is said to be a *saddle point*.

(b) Equal eigenvalues

If $\boldsymbol{J}$ is diagonal, the canonical system has solutions given by (2.44) with $\lambda_1 = \lambda_2 = \lambda_0 \neq 0$. Thus (2.18(b)) corresponds to a special node, called a *star* (stable if $\lambda_0 < 0$; unstable if $\lambda_0 > 0$), in which the non-trivial trajectories are all radial straight lines (see Fig. 2.3).

If $\boldsymbol{J}$ is not diagonal (i.e. (2.18(c))) then we must consider

$$\dot{y}_1 = \lambda_0 y_1 + y_2, \qquad \dot{y}_2 = \lambda_0 y_2; \qquad \lambda_0 \neq 0. \tag{2.47}$$

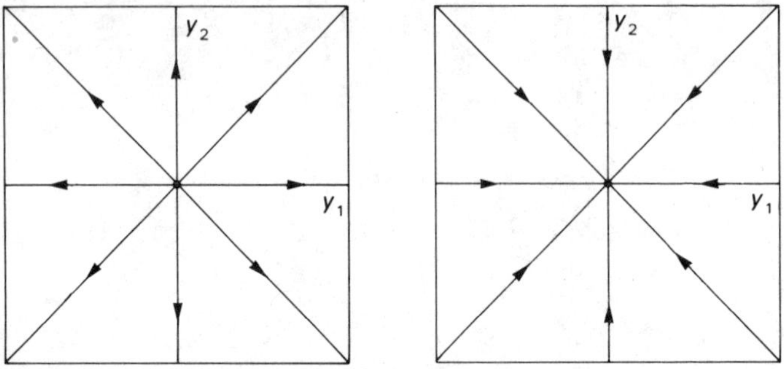

(a) (b)

Fig. 2.3. Equal eigenvalues ($\lambda_1 = \lambda_2 = \lambda_0$) give rise to star nodes: (a) unstable; (b) stable; when $\boldsymbol{A}$ is diagonal.

This system has solutions

$$y_1(t) = (C_1 + tC_2)e^{\lambda_0 t}; \qquad y_2(t) = C_2 e^{\lambda_0 t} \tag{2.48}$$

and phase portraits like those in Fig. 2.4 are obtained (see Example 1.4.2).

The origin is said to be an *improper node* (stable $\lambda_0 < 0$; unstable $\lambda_0 > 0$). The line on which the trajectories change direction is the locus of extreme values for y_1. This is given by the $\dot{y}_1 = 0$ isocline, namely

$$y_2 = -\lambda_0 y_1. \tag{2.49}$$

(c) Complex eigenvalues

The Jordan matrix is given by (2.18(d)) so the canonical system is

$$\dot{y}_1 = \alpha y_1 - \beta y_2; \qquad \dot{y}_2 = \beta y_1 + \alpha y_2. \tag{2.50}$$

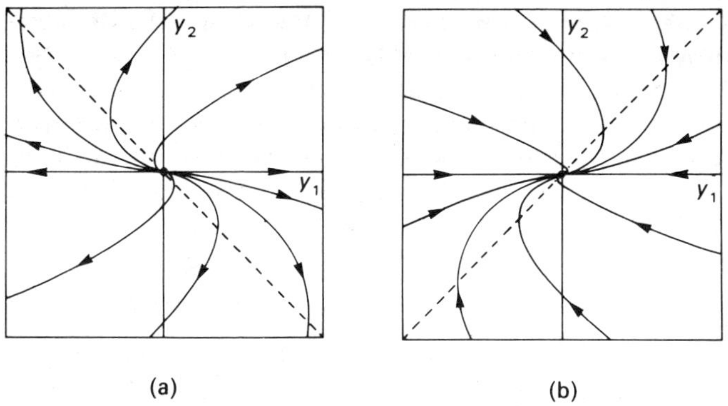

Fig. 2.4. When A is not diagonal, equal eigenvalues indicate that the origin is an improper node: (a) unstable ($\lambda_0 > 0$); (b) stable ($\lambda_0 < 0$).

This kind of system can be solved by introducing plane polar coordinates such that $y_1 = r\cos\theta$, $y_2 = r\sin\theta$. We obtain, as in Example 1.4.1,

$$\dot{r} = \alpha r, \qquad \dot{\theta} = \beta \tag{2.51}$$

with solutions

$$r(t) = r_0 e^{\alpha t}, \qquad \theta(t) = \beta t + \theta_0. \tag{2.52}$$

Typical phase portraits are shown in Fig. 2.5.

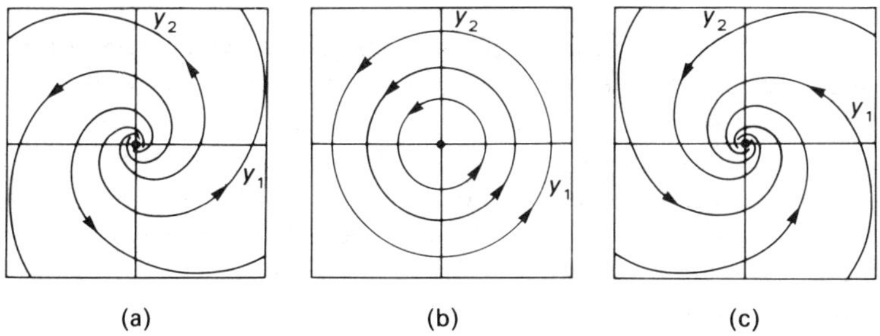

Fig. 2.5. Complex eigenvalues give rise to (a) unstable foci ($\alpha > 0$), (b) centres ($\alpha = 0$) and (c) stable foci ($\alpha < 0$).

If $\alpha \neq 0$ the origin is said to be a *focus* (stable if $\alpha < 0$; unstable if $\alpha > 0$). The phase portrait is often said to consist of an *attracting*

($\alpha < 0$) or *repelling* ($\alpha > 0$) *spiral*. The parameter $\beta > 0$ determines the angular speed of description of the spiral.

When $\alpha = 0$, the origin is said to be a *centre* and the phase portrait consists of a continuum of concentric circles. This is the only non-trivial way in which *recurrent* or *periodic* behaviour occurs in linear systems. Every point (excluding the origin) in the phase plane recurs at intervals of $T = 2\pi/\beta$. The coordinates are periodic in t with this period, i.e.,

$$y_1 = r_0 \cos(\beta t + \theta_0), \qquad y_2 = r_0 \sin(\beta t + \theta_0). \tag{2.53}$$

The other (trivial) form of recurrence that occurs in simple linear systems is the fixed point. In a sense, the fixed point is the ultimate in recurrence; it recurs instantaneously with period zero.

2.3.2 *Non-simple canonical systems*

A linear system $\dot{\boldsymbol{x}} = \boldsymbol{A}\boldsymbol{x}$ is non-simple if $\boldsymbol{A}$ is singular (i.e. $\det \boldsymbol{A} = 0$ and at least one of the eigenvalues of $\boldsymbol{A}$ is zero). It follows that there are non-trivial solutions to $\boldsymbol{A}\boldsymbol{x} = \boldsymbol{0}$ and the system has fixed points other than $\boldsymbol{x} = \boldsymbol{0}$. For linear systems in the plane, there are only two possibilities: either the rank of $\boldsymbol{A}$ is one; or $\boldsymbol{A}$ is null. In the first case there is a *line* of fixed points passing through the origin; in the second, every point in the plane is a fixed point. Of course, the rank of $\boldsymbol{J}$ is equal to the rank of $\boldsymbol{A}$, so that the canonical systems exhibit corresponding non-simple behaviour. Figure 2.6 shows two examples when $\boldsymbol{J}$ has rank one.

We shall, in the remainder of this chapter, focus attention on simple linear systems. These systems play a key role in understanding the nature of the fixed points in non-linear systems (see Section 3.3).

The results obtained in this section are summarized in Fig. 2.7; where each type of phase portrait is associated with a set of points in the $\operatorname{tr}(\boldsymbol{A})$–$\det(\boldsymbol{A})$ plane. Each point of this plane corresponds to a particular pair of eigenvalues for $\boldsymbol{A}$ and therefore to a particular canonical system.

2.4 Classification of simple linear phase portraits in the plane

2.4.1 *Phase portrait of a simple linear system*

The phase portrait of any linear system $\dot{\boldsymbol{x}} = \boldsymbol{A}\boldsymbol{x}$ can be obtained from that of its canonical form $\dot{\boldsymbol{y}} = \boldsymbol{J}\boldsymbol{y}$ by applying the transformation

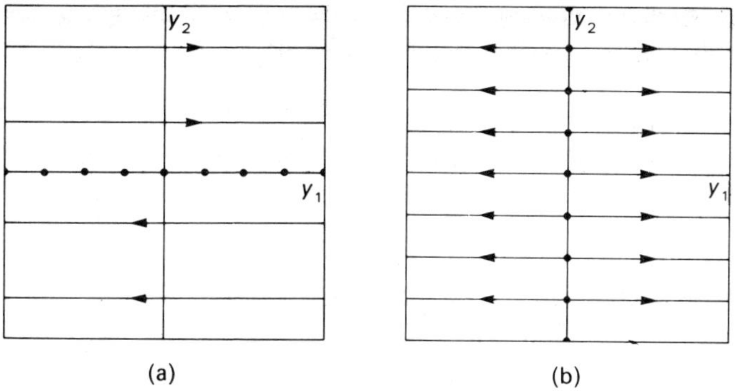

Fig. 2.6. (a)

$$\begin{bmatrix} \dot{y}_1 \\ \dot{y}_2 \end{bmatrix} = \begin{bmatrix} 0 & 1 \\ 0 & 0 \end{bmatrix} \begin{bmatrix} y_1 \\ y_2 \end{bmatrix}, \quad \text{i.e. } \lambda_0 = 0.$$

Every point on the y_1-axis is a fixed point; cf. the $\lambda_0 \to 0$ limit of Fig. 2.4(a).

(b)

$$\begin{bmatrix} \dot{y}_1 \\ \dot{y}_2 \end{bmatrix} = \begin{bmatrix} 1 & 0 \\ 0 & 0 \end{bmatrix} \begin{bmatrix} y_1 \\ y_2 \end{bmatrix}, \quad \text{i.e. } \lambda_1 = 1, \lambda_2 = 0.$$

Every point on the y_2-axis is a fixed point; cf. the $\lambda_2 \to 0$ limit of Fig. 2.1(a).

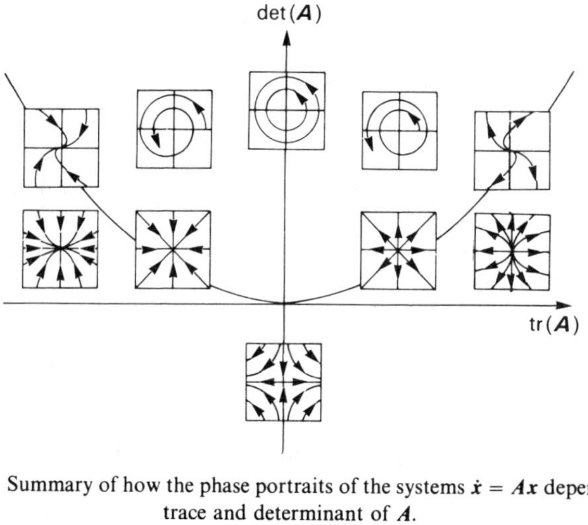

Fig. 2.7. Summary of how the phase portraits of the systems $\dot{\boldsymbol{x}} = \boldsymbol{A}\boldsymbol{x}$ depend on the trace and determinant of $\boldsymbol{A}$.

$\boldsymbol{x} = \boldsymbol{M}\boldsymbol{y}$. Let us consider how the phase portrait is changed under this transformation. For example, the canonical system

$$\dot{y}_1 = y_1; \qquad \dot{y}_2 = -y_2 \tag{2.54}$$

has the phase portrait shown in Fig. 2.8(a). Figure 2.8(b)–(f) shows the phase portraits for some linear systems that have (2.54) as their canonical system.

As noted in (2.6) the variables y_1 and y_2 are the coordinates of $\mathbf{x}$ relative to the basis $\{\boldsymbol{m}_1, \boldsymbol{m}_2\}$ obtained from the columns of $\boldsymbol{M}$. Thus the y_1- and y_2-axes are represented by straight lines, through the origin of the x_1x_2-plane, in the directions of $\boldsymbol{m}_1$ and $\boldsymbol{m}_2$ respectively.

The directions defined by $\boldsymbol{m}_1$ and $\boldsymbol{m}_2$ are called the *principal directions* at the origin. The mapping is a bijection so every point of the y_1y_2-plane is uniquely represented in the x_1x_2-plane and vice versa. Furthermore, $\boldsymbol{x}$ is a continuous function of $\boldsymbol{y}$, so that trajectories map onto trajectories. Orientation of the trajectories is also preserved. In particular, the directions of the separatrices in the x_1x_2-plane are given by the eigenvectors of $\boldsymbol{A}$.

These properties of the linear transformation $\boldsymbol{x} = \boldsymbol{M}\boldsymbol{y}$ are sufficient to ensure that, although the phase portrait is distorted, the origin is still a saddle point. The phase portraits in Fig. 2.8(a)–(f) are typical of *all* systems $\dot{\boldsymbol{x}} = \boldsymbol{A}\boldsymbol{x}$ for which $\boldsymbol{A}$ has eigenvalues $+1$ and -1; they all have saddle points at the origin. In other words, linear transformations preserve the qualitative behaviour of the solutions.

Example 2.4.1. Sketch the phase portrait of the system

$$\dot{y}_1 = 2y_1, \qquad \dot{y}_2 = y_2, \tag{2.55}$$

and the corresponding phase portraits in the x_1x_2-plane where

$$x_1 = y_1 + y_2, \qquad x_2 = y_1 + 2y_2, \tag{2.56}$$

and

$$x_1 = y_1, \qquad x_2 = -y_1 + y_2 \tag{2.57}$$

respectively.

Solution. The phase portrait of system (2.55) is sketched in Fig. 2.9(a).

Observe that the change of variables given by (2.56) and (2.57) can

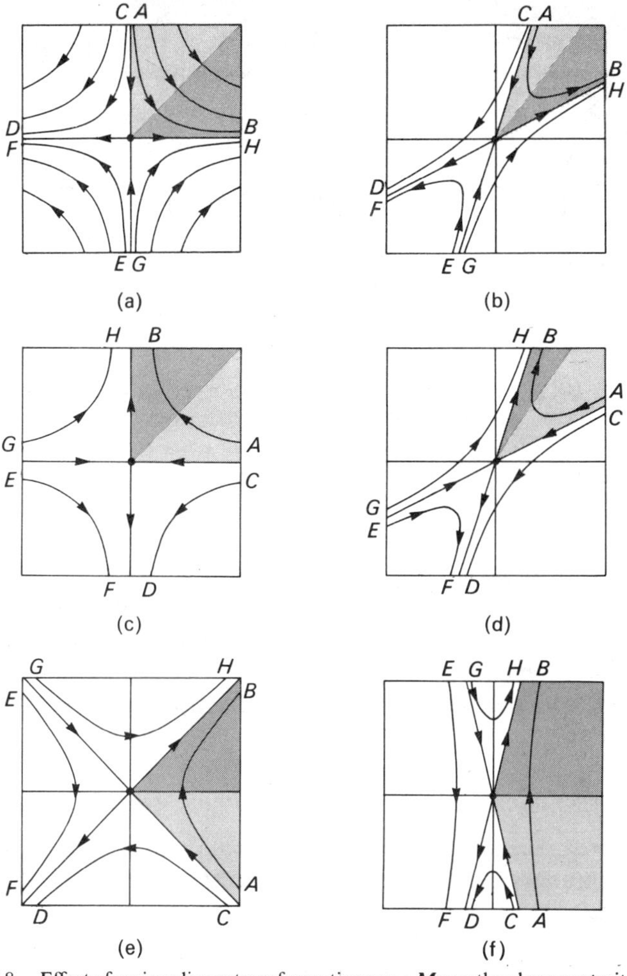

Fig. 2.8. Effect of various linear transformations $\boldsymbol{x} = \boldsymbol{M}\boldsymbol{y}$ on the phase portrait of the canonical system $\dot{y}_1 = y_1$, $\dot{y}_2 = -y_2$.

(a) Canonical system $\dot{y}_1 = y_1$; $\dot{y}_2 = -y_2$.

(b) $\boldsymbol{M} = \begin{bmatrix} 2 & 1 \\ 1 & 3 \end{bmatrix}$. (c) $\boldsymbol{M} = \begin{bmatrix} 0 & 1 \\ 1 & 0 \end{bmatrix}$.

(d) $\boldsymbol{M} = \begin{bmatrix} 1 & 2 \\ 3 & 1 \end{bmatrix}$. (e) $\boldsymbol{M} = \begin{bmatrix} 1 & 1 \\ 1 & -1 \end{bmatrix}$.

(f) $\boldsymbol{M} = \begin{bmatrix} 1 & 1 \\ 4 & -4 \end{bmatrix}$.

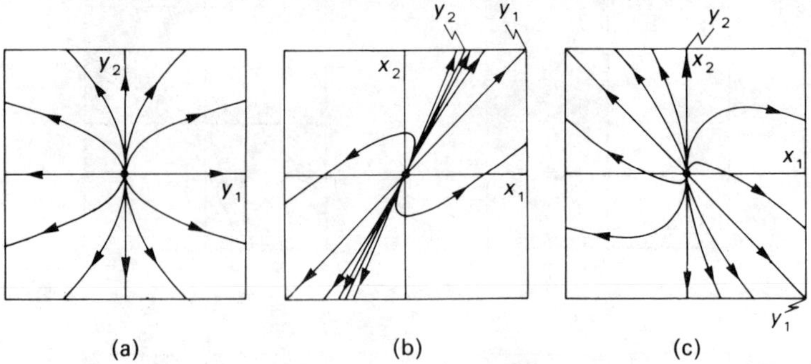

Fig. 2.9. The phase portraits of the canonical system (2.55) in the y_1, y_2-plane [(a)] and in the x_1, x_2-plane defined by (2.56) and (2.57) [(b) and (c) respectively].

be written in the matrix form $\boldsymbol{x} = \boldsymbol{M}\boldsymbol{y}$ with

$$\boldsymbol{M} = \begin{bmatrix} 1 & 1 \\ 1 & 2 \end{bmatrix} \quad \text{and} \quad \begin{bmatrix} 1 & 0 \\ -1 & 1 \end{bmatrix}$$

respectively. The basis for which y_1 and y_2 are coordinates is $\{(1, 1), (1, 2)\}$ for (2.56) and $\{(1, -1), (0, 1)\}$ for (2.57). The y_1 and y_2-coordinate directions are shown in the x_1x_2-plane in Fig. 2.9(b) and (c) along with the correspondingly transformed phase portrait from Fig. 2.9(a). □

2.4.2 *Types of canonical system and qualitative equivalence*

Two Jordan matrices with different eigenvalues are not similar and so their solutions cannot be related by a linear transformation. However, the fixed point at the origin of every simple canonical system is one of a small number of possibilities; namely a node, a focus, a saddle, etc. How are these phase portraits related?

Consider, for example, the canonical systems

$$\dot{y}_1 = \alpha y_1 - \beta y_2, \qquad \dot{y}_2 = \beta y_1 + \alpha y_2, \tag{2.58}$$

$\alpha, \beta > 0$ and

$$\dot{z}_1 = z_1 - z_2, \qquad \dot{z}_2 = z_1 + z_2. \tag{2.59}$$

Their phase portraits are both unstable foci (see Fig. 2.10).

The trajectories are given by

$$r = r_0 e^{\alpha t}, \qquad \theta = \beta t + \theta_0 \tag{2.60}$$

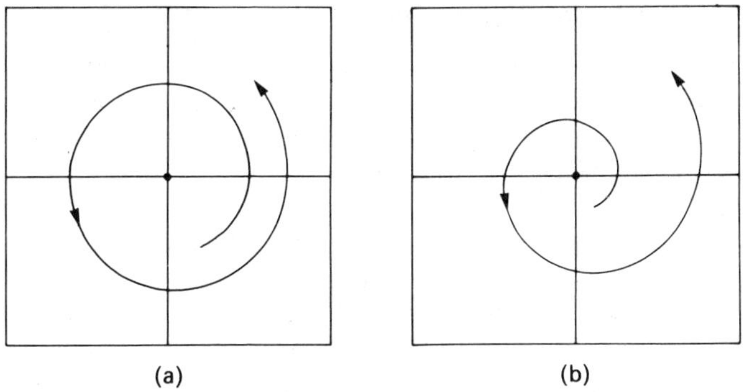

Fig. 2.10. Relation between phase portraits of canonical systems: (a) $y_1 y_2$-plane: trajectory through (r_0, θ_0) of the system (2.60); (b) $z_1 z_2$-plane: trajectory through (R_0, ϕ_0) of the system (2.61).

and

$$R = R_0 e^t, \qquad \phi = t + \phi_0, \tag{2.61}$$

respectively, where (r, θ) and (R, ϕ) are polar coordinates in the $y_1 y_2$- and $z_1 z_2$-planes. The trajectory passing through (r_0, θ_0), $r_0 \neq 0$, in the y_1, y_2-plane can be mapped onto the trajectory passing through (R_0, ϕ_0) in the $z_1 z_2$-plane by the transformation

$$R = R_0 \left(\frac{r}{r_0}\right)^{1/\alpha}, \qquad \phi = \frac{\theta - \theta_0}{\beta} + \phi_0. \tag{2.62}$$

This transformation allows the phase portrait in Fig. 2.10 (a) to be mapped – trajectory by trajectory – onto that in Fig. 2.10 (b). The transformation in (2.62) is a bijection; it is continuous; it preserves the orientation of the trajectories. However, it does *not* map the $y_1 y_2$-plane onto the $z_1 z_2$-plane in a *linear* way, unless $\alpha = \beta = 1$.

In Sections 2.4.1 and 2.4.2 we have examined more closely the relationship between systems with the same qualitative behaviour (see also Exercise 2.17). We are led to make the intuitive ideas of Section 1.3 more precise in the following definition.

Definition 2.4.1. Two systems of first-order differential equations are said to be *qualitatively equivalent* if there is a continuous bijection which maps the phase portrait of one onto that of the other in such a way as to preserve the orientation of the trajectories.

2.4.3 *Classification of linear systems*

In this chapter, we have illustrated the fact that every linear system in the plane is qualitatively equivalent to one of the systems whose phase portraits are shown in Fig. 2.11. The ten phase portraits shown are representative of the *algebraic type* of the linear system.

Qualitative equivalence goes further than this. It can be shown that all stable (unstable) nodes, improper nodes and foci are also related in the manner described in Definition 2.4.1. This means that the algebraic types can be further grouped into *qualitative* (or topological) *types* as indicated in Fig. 2.11. In these terms, simple linear systems only exhibit four kinds of qualitative behaviour: stable, centre, saddle and unstable.

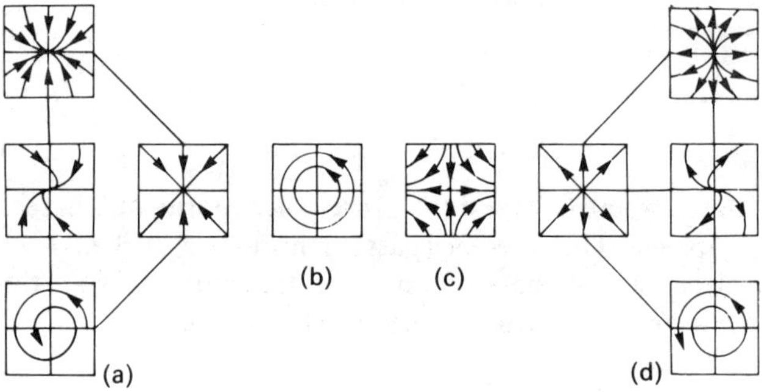

Fig. 2.11. The distinct qualitative types of phase portrait for linear systems: (a) stable; (b) centre; (c) saddle; (d) unstable; in relation to the algebraic types obtained in Section 2.3.

2.5 The evolution operator

In Section 1.5 we suggested that the trajectories of linear systems in the plane could be described by an evolution matrix. We will now consider this idea in more detail.

For any real $n \times n$ matrix $\boldsymbol{P}$ define the *exponential matrix* $e^{\boldsymbol{P}}$ by

$$e^{\boldsymbol{P}} = \sum_{k=0}^{\infty} \frac{\boldsymbol{P}^k}{k!} \tag{2.63}$$

with $\boldsymbol{P}^0 = \boldsymbol{I}_n$, the $n \times n$ unit matrix. Let $\boldsymbol{Q}$ be a real $n \times n$ matrix such that $\boldsymbol{PQ} = \boldsymbol{QP}$ then it follows from (2.63) that

$$e^{\boldsymbol{P}+\boldsymbol{Q}} = e^{\boldsymbol{P}}e^{\boldsymbol{Q}}, \tag{2.64}$$

(see Exercise 2.24). This result allows us to conclude that

$$e^{P}e^{-P} = e^{-P}e^{P} = e^{O} = \boldsymbol{I}_n, \tag{2.65}$$

so that

$$(e^{P})^{-1} = e^{-P}. \tag{2.66}$$

If we let $\boldsymbol{P} = \boldsymbol{A}t$ and differentiate (2.63) with respect to t, we find

$$\frac{d}{dt}(e^{At}) = Ae^{At} = e^{At}A. \tag{2.67}$$

This means that

$$\frac{d}{dt}(e^{-At}\boldsymbol{x}) = e^{-At}\dot{\boldsymbol{x}} - e^{-At}\boldsymbol{A}\boldsymbol{x} = e^{-At}(\dot{\boldsymbol{x}} - \boldsymbol{A}\boldsymbol{x}) \equiv \boldsymbol{0}. \tag{2.68}$$

If we are given $\boldsymbol{x}(t_0) = \boldsymbol{x}_0$, then integrating (2.68) from t_0 to t, gives

$$e^{-At}\boldsymbol{x}(t) = e^{-At_0}\boldsymbol{x}_0 \tag{2.69}$$

and hence

$$\boldsymbol{x}(t) = e^{A(t-t_0)}\boldsymbol{x}_0 = \phi_{t-t_0}(\boldsymbol{x}_0), \tag{2.70}$$

by (2.66). Thus, matrix multiplication by $e^{A(t-t_0)}$ gives the evolution of the phase point which is at $\boldsymbol{x}_0$ when $t = t_0$. It follows that e^{At} is the evolution operator for the linear system $\dot{\boldsymbol{x}} = \boldsymbol{A}\boldsymbol{x}$.

Of course, in order to make use of (2.70) we must know $e^{A(t-t_0)}$. For simple canonical systems in the plane we can write down e^{Jt} from the solutions given in Section 2.3.1. We find:

$$e^{Jt} = \begin{bmatrix} e^{\lambda_1 t} & 0 \\ 0 & e^{\lambda_2 t} \end{bmatrix} \quad \text{for } \boldsymbol{J} = \begin{bmatrix} \lambda_1 & 0 \\ 0 & \lambda_2 \end{bmatrix} \tag{2.71}$$

(including $\lambda_1 = \lambda_2 = \lambda_0$) from (2.44);

$$e^{Jt} = e^{\lambda_0 t}\begin{bmatrix} 1 & t \\ 0 & 1 \end{bmatrix} \quad \text{for } \boldsymbol{J} = \begin{bmatrix} \lambda_0 & 1 \\ 0 & \lambda_0 \end{bmatrix} \tag{2.72}$$

from (2.48); and

$$e^{Jt} = e^{\alpha t}\begin{bmatrix} \cos\beta t & -\sin\beta t \\ \sin\beta t & \cos\beta t \end{bmatrix} \quad \text{for } \boldsymbol{J} = \begin{bmatrix} \alpha & -\beta \\ \beta & \alpha \end{bmatrix} \tag{2.73}$$

from (2.52). These results can also be obtained directly from (2.63) (see Exercise 2.25).

Equations (2.71)–(2.73) can be used to obtain e^{At} for any non-singular 2×2 matrix $\boldsymbol{A}$ by exploiting the relation between $\boldsymbol{A}$ and its Jordan form $\boldsymbol{J}$. Recall,

$$\boldsymbol{x}(t) = \boldsymbol{M}\boldsymbol{y}(t) = \boldsymbol{M}e^{\boldsymbol{J}t}\boldsymbol{y}(0) = (\boldsymbol{M}e^{\boldsymbol{J}t}\boldsymbol{M}^{-1})\boldsymbol{x}(0), \tag{2.74}$$

so that

$$e^{\boldsymbol{A}t} = \boldsymbol{M}e^{\boldsymbol{J}t}\boldsymbol{M}^{-1}. \tag{2.75}$$

While this gives us an understandable way of obtaining e^{At} it is not always convenient in practice. An alternative way is to adapt a general method of Sylvester (Barnett, 1975) for finding functions of matrices. We consider the method for $n = 2$.

Let the 2×2 real matrix $\boldsymbol{A}$ have two distinct real or complex eigenvalues λ_1, λ_2. Define the (possibly complex) matrices $\boldsymbol{Q}_1, \boldsymbol{Q}_2$ by

$$\boldsymbol{Q}_1 = \frac{\boldsymbol{A} - \lambda_2 \boldsymbol{I}}{\lambda_1 - \lambda_2}, \qquad \boldsymbol{Q}_2 = \frac{\boldsymbol{A} - \lambda_1 \boldsymbol{I}}{\lambda_2 - \lambda_1} \tag{2.76}$$

and observe that $\boldsymbol{A} = \lambda_1 \boldsymbol{Q}_1 + \lambda_2 \boldsymbol{Q}_2$. It can be shown that $\boldsymbol{Q}_1\boldsymbol{Q}_2 = \boldsymbol{Q}_2\boldsymbol{Q}_1 = \boldsymbol{0}$, $\boldsymbol{Q}_1^2 = \boldsymbol{Q}_1$ and $\boldsymbol{Q}_2^2 = \boldsymbol{Q}_2$ (see Exercise 2.26) and therefore

$$\boldsymbol{A}^k = (\lambda_1 \boldsymbol{Q}_1 + \lambda_2 \boldsymbol{Q}_2)^k = \lambda_1^k \boldsymbol{Q}_1 + \lambda_2^k \boldsymbol{Q}_2 \tag{2.77}$$

for any positive integer k. Thus

$$\begin{aligned} e^{\boldsymbol{A}t} &= \sum_{k=0}^{\infty} \frac{\boldsymbol{A}^k t^k}{k!} = \sum_{k=0}^{\infty} \left(\frac{\lambda_1^k t^k}{k!} \boldsymbol{Q}_1 + \frac{\lambda_2^k t^k}{k!} \boldsymbol{Q}_2 \right) \\ &= e^{\lambda_1 t} \boldsymbol{Q}_1 + e^{\lambda_2 t} \boldsymbol{Q}_2. \end{aligned} \tag{2.78}$$

When the eigenvalues λ_1, λ_2 of $\boldsymbol{A}$ are equal, define $\boldsymbol{Q} = \boldsymbol{A} - \lambda_0 \boldsymbol{I}$, where $\lambda_1 = \lambda_2 = \lambda_0$. In this case $\boldsymbol{Q}^2 = \boldsymbol{0}$ and

$$\boldsymbol{A}^k = (\lambda_0 \boldsymbol{I} + \boldsymbol{Q})^k = \lambda_0^k \boldsymbol{I} + k\lambda_0^{k-1} \boldsymbol{Q}, \tag{2.79}$$

because $\boldsymbol{Q}^k = \boldsymbol{0}$ for $k \geqslant 2$. Thus

$$\begin{aligned} e^{\boldsymbol{A}t} &= \sum_{k=0}^{\infty} \left(\frac{\lambda_0^k}{k!} \boldsymbol{I} + \frac{k\lambda_0^{k-1}}{k!} \boldsymbol{Q} \right) t^k \\ &= e^{\lambda_0 t} (\boldsymbol{I} + t(\boldsymbol{A} - \lambda_0 \boldsymbol{I})). \end{aligned} \tag{2.80}$$

Example 2.5.1. Find e^{At} where

$$A = \begin{bmatrix} 1 & 1 \\ -1 & 3 \end{bmatrix} \tag{2.81}$$

by (a) the method of reducing to canonical form, and (b) the use of formula (2.78) or (2.80).

Solution. (a) The eigenvalues of A are both $\lambda_0 = 2$. An eigenvector u_0 is given by $u_0 = \begin{bmatrix} 1 \\ 1 \end{bmatrix}$. Let

$$M = \begin{bmatrix} 1 & -1 \\ 1 & 0 \end{bmatrix}, \tag{2.82}$$

then

$$M^{-1}AM = \begin{bmatrix} 2 & 1 \\ 0 & 2 \end{bmatrix} = J; \tag{2.83}$$

the required Jordan matrix. The exponential matrix

$$e^{Jt} = e^{2t}\begin{bmatrix} 1 & t \\ 0 & 1 \end{bmatrix}$$

by (2.72), and (2.75) gives

$$\begin{aligned} e^{At} = Me^{Jt}M^{-1} &= e^{2t}\begin{bmatrix} 1 & -1 \\ 1 & 0 \end{bmatrix}\begin{bmatrix} 1 & t \\ 0 & 1 \end{bmatrix}\begin{bmatrix} 0 & 1 \\ -1 & 1 \end{bmatrix} \\ &= e^{2t}\begin{bmatrix} 1-t & t \\ -t & 1+t \end{bmatrix}. \end{aligned} \tag{2.84}$$

(b) The eigenvalues of A have been calculated in (a). Substituting for A and $\lambda_0 = 2$ in (2.80), we obtain

$$\begin{aligned} e^{At} &= e^{2t}\left\{I + t\left\{\begin{bmatrix} 1 & 1 \\ -1 & 3 \end{bmatrix} - 2\begin{bmatrix} 1 & 0 \\ 0 & 1 \end{bmatrix}\right\}\right\} \\ &= e^{2t}\begin{bmatrix} 1-t & t \\ -t & 1+t \end{bmatrix}. \end{aligned} \tag{2.85}$$

□

2.6 Affine systems

In this section we consider systems where the vector field **X** is of the form

$$X(x) = Ax + h \tag{2.86}$$

where $h \in R^n$. The mapping X is said to be *affine*. In the strict algebraic sense $Ax + h$ is a *non-linear* function of x but it is clearly closely related to linear functions. In fact, systems $\dot{x} = X(x)$ with X given by (2.86) are frequently called *non-homogeneous linear* systems as opposed to the *homogeneous linear* system $\dot{x} = Ax$.

The change of variable

$$x = y - x_0, \tag{2.87}$$

where x_0 is a solution of $Ax_0 = h$, if one exists; allows (2.86) to be written as

$$Y(y) = X(y - x_0) = A(y - x_0) + h = Ay. \tag{2.88}$$

In such cases, the change of variable (2.87) allows the affine system

$$\dot{x} = Ax + h \tag{2.89}$$

to be written as the linear system

$$\dot{y} = Ay. \tag{2.90}$$

However, what if h is not in the image of A or if h in (2.89) is replaced by a time dependent vector $h(t)$? Non-autonomous systems of the latter form frequently occur in practical problems (see Section 4.2). The evolution operator provides a solution to both of these difficulties.

The procedure is a matrix equivalent of the 'integrating factor method' used in Exercise 1.1. Multiply (2.89) throughout by e^{-At} and rearrange as

$$\frac{d}{dt}(e^{-At}x) = e^{-At}h(t). \tag{2.91}$$

Here the evolution operator acts as a 'matrix integrating factor'. Suppose we are given $x(t_0) = x_0$, then integrating (2.91) from t_0 to t gives

$$e^{-At}x(t) - e^{-At_0}x_0 = \int_{t_0}^{t} e^{-As}h(s)\,ds.$$

Since $e^{At}e^{-At} = I$ we can solve for $x(t)$; thus

$$x(t) = x_{CF}(t) + x_{PI}(t), \tag{2.92}$$

where

$$\boldsymbol{x}_{\mathrm{CF}}(t) = \mathrm{e}^{\boldsymbol{A}(t-t_0)}\boldsymbol{x}_0 \tag{2.93}$$

and

$$\boldsymbol{x}_{\mathrm{PI}}(t) = \mathrm{e}^{\boldsymbol{A}t}\int_{t_0}^{t} \mathrm{e}^{-\boldsymbol{A}s}\boldsymbol{h}(s)\,\mathrm{d}s. \tag{2.94}$$

Equation (2.92) gives the general solution to the affine system (2.89), whether or not $\boldsymbol{h}$ depends on t.

The first term $\boldsymbol{x}_{\mathrm{CF}}(t)$ is the solution of the linear system $\dot{\boldsymbol{x}} = \boldsymbol{A}\boldsymbol{x}$ which satisfies the initial condition $\boldsymbol{x}_{\mathrm{CF}}(t_0) = \boldsymbol{x}_0$. It corresponds to the *complementary function* of traditional treatments of second-order, non-homogeneous linear equations. The second term, $\boldsymbol{x}_{\mathrm{PI}}(t)$, gives a general expression for the '*particular integral*' of such treatments. It is easy to verify that $\boldsymbol{x}_{\mathrm{PI}}(t)$ satisfies (2.89) and that $\boldsymbol{x}_{\mathrm{PI}}(0) = \boldsymbol{0}$, so that $\boldsymbol{x}(t)$ in (2.92) satisfies both (2.89) and $\boldsymbol{x}(t_0) = \boldsymbol{x}_0$.

Example 2.6.1. Consider the second-order equation

$$\ddot{x} + 2\dot{x} + 2x = u(t). \tag{2.95}$$

Write this equation as a first-order system and obtain a general expression for the solution which satisfies $x(0) = \dot{x}(0) = 0$.

Solution. Let $x_1 = x$ and $x_2 = \dot{x}$, then (2.95) becomes

$$\dot{x}_1 = x_2, \qquad \dot{x}_2 = -2x_1 - 2x_2 + u(t). \tag{2.96}$$

In matrix notation, (2.96) takes the form $\dot{\boldsymbol{x}} = \boldsymbol{A}\boldsymbol{x} + \boldsymbol{h}(t)$ with

$$\boldsymbol{A} = \begin{bmatrix} 0 & 1 \\ -2 & -2 \end{bmatrix} \quad \text{and} \quad \boldsymbol{h}(t) = \begin{bmatrix} 0 \\ u(t) \end{bmatrix}. \tag{2.97}$$

The matrix $\boldsymbol{A}$ has eigenvalues $\lambda_1 = -1 + i$, $\lambda_2 = -1 - i$ and (2.78) gives

$$\mathrm{e}^{\boldsymbol{A}t} = \mathrm{e}^{-t}\left\{\frac{\mathrm{e}^{it}}{2i}\begin{bmatrix} 1+i & 1 \\ -2 & -1+i \end{bmatrix} - \frac{\mathrm{e}^{-it}}{2i}\begin{bmatrix} 1-i & 1 \\ -2 & -1-i \end{bmatrix}\right\}$$

$$\mathrm{e}^{\boldsymbol{A}t} = \mathrm{e}^{-t}\begin{bmatrix} \cos t + \sin t & \sin t \\ -2\sin t & \cos t - \sin t \end{bmatrix}. \tag{2.98}$$

The initial condition given corresponds to $\boldsymbol{x}(0) = \boldsymbol{0}$, so $\boldsymbol{x}_{\mathrm{CF}}(t) \equiv \boldsymbol{0}$ in

(2.92) and $\boldsymbol{x}(t) = \boldsymbol{x}_{\mathrm{PI}}(t)$, which is

$$\mathrm{e}^{-t}\begin{bmatrix} \cos t + \sin t & \sin t \\ -2\sin t & \cos t - \sin t \end{bmatrix} \int_0^t \mathrm{e}^s u(s) \begin{bmatrix} -\sin s \\ \cos s + \sin s \end{bmatrix} \mathrm{d}s. \quad (2.99)$$

Hence

$$x_1(t) = x(t) = \mathrm{e}^{-t}\left\{ \sin t \int_0^t \mathrm{e}^s u(s) \cos s \, \mathrm{d}s - \cos t \int_0^t \mathrm{e}^s u(s) \sin s \, \mathrm{d}s \right\}. \quad (2.100)\ \square$$

2.7 Linear systems of dimension greater than two

2.7.1 *Three-dimensional systems*

The real Jordan forms for 3×3 matrices are

$$\begin{bmatrix} \lambda_0 & 0 & 0 \\ 0 & \lambda_1 & 0 \\ 0 & 0 & \lambda_2 \end{bmatrix}, \quad \begin{bmatrix} \alpha & -\beta & 0 \\ \beta & \alpha & 0 \\ 0 & 0 & \lambda_1 \end{bmatrix},$$

$$\begin{bmatrix} \lambda_0 & 1 & 0 \\ 0 & \lambda_0 & 0 \\ 0 & 0 & \lambda_1 \end{bmatrix} \text{ and } \begin{bmatrix} \lambda_0 & 1 & 0 \\ 0 & \lambda_0 & 1 \\ 0 & 0 & \lambda_0 \end{bmatrix}, \quad (2.101)$$

where α, β, λ_0, λ_1, λ_2 are real (Hirsch and Smale, 1974). Observe that all, except the last form, in (2.101) can be partitioned in diagonal blocks of dimension 1 and 2. For example, consider

$$\dot{\boldsymbol{y}} = \boldsymbol{J}\boldsymbol{y} \quad \text{with} \quad \boldsymbol{J} = \left[\begin{array}{cc|c} \alpha & -\beta & 0 \\ \beta & \alpha & 0 \\ \hline 0 & 0 & \lambda_1 \end{array}\right] \quad (2.102)$$

where $\lambda_1 < \alpha < 0$, $\beta > 0$. The diagonal block structure ensures that the y_3-equation is decoupled from the other two. In fact, (2.102) is equivalent to the two systems

$$\begin{bmatrix} \dot{y}_1 \\ \dot{y}_2 \end{bmatrix} = \begin{bmatrix} \alpha & -\beta \\ \beta & \alpha \end{bmatrix} \begin{bmatrix} y_1 \\ y_2 \end{bmatrix}, \quad (2.103)$$

$$\dot{y}_3 = \lambda_1 y_3. \quad (2.104)$$

This decoupling allows us to construct the qualitative features of the three-dimensional phase portrait.

Equation (2.103) shows that the projection of the trajectories onto a plane of constant y_3 is a stable ($\alpha < 0$) focus; the attracting spiral being described in an anticlockwise sense ($\beta > 0$). Since $\lambda_1 < 0, |y_3(t)|$ decreases exponentially with increasing t. A typical trajectory of (2.102) passing through (a, b, c) at $t = 0$, with all three coordinates non-zero, is shown in Fig. 2.12. The spiral in the constant y_3 projection is drawn out into a kind of helix with decreasing radius and pitch.

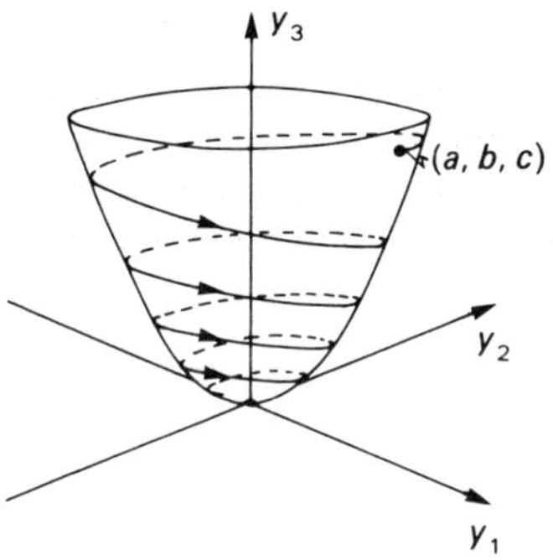

Fig. 2.12. The trajectory of (2.102) passing through (a, b, c). Each trajectory lies in a surface of the form $y_3 = K\,(y_1^2 + y_2^2)^{\lambda_1/2\alpha}$.

Returning to (2.101), the Jordan form that is not amenable to block decomposition corresponds to a system $\dot{\boldsymbol{y}} = \boldsymbol{J}\boldsymbol{y}$ that can be solved by finding $e^{\boldsymbol{J}t}$. The evolution operator can be obtained by using the expansion (2.63); observe

$$\boldsymbol{J} = \begin{bmatrix} \lambda_0 & 1 & 0 \\ 0 & \lambda_0 & 1 \\ 0 & 0 & \lambda_0 \end{bmatrix} = \begin{bmatrix} \lambda_0 & 0 & 0 \\ 0 & \lambda_0 & 0 \\ 0 & 0 & \lambda_0 \end{bmatrix} + \begin{bmatrix} 0 & 1 & 0 \\ 0 & 0 & 1 \\ 0 & 0 & 0 \end{bmatrix}$$

$$\boldsymbol{J} = \boldsymbol{D} + \boldsymbol{C}. \tag{2.105}$$

As in Exercise 2.24, $\boldsymbol{DC} = \boldsymbol{CD}$ and (2.64) reduces the problem to one of finding $e^{\boldsymbol{C}t}$. In this case,

$$\boldsymbol{C}^3 = \boldsymbol{0} \quad \text{and} \quad e^{\boldsymbol{C}t} = \boldsymbol{I} + t\boldsymbol{C} + \tfrac{1}{2}t^2\boldsymbol{C}^2$$

$$\text{and} \quad e^{\boldsymbol{J}t} = e^{\lambda_0 t}\begin{bmatrix} 1 & t & \tfrac{1}{2}t^2 \\ 0 & 1 & t \\ 0 & 0 & 1 \end{bmatrix}. \tag{2.106}$$

Example 2.7.1. Sketch the trajectory which satisfies $\mathbf{x}(0) = (0, b, c)$; $b, c > 0$, for the system

$$\dot{\mathbf{x}} = \boldsymbol{J}\mathbf{x}, \tag{2.107}$$

where $\boldsymbol{J}$ is given by (2.105) with $\lambda_0 > 0$.

Solution. Observe that the projection of (2.107) onto the x_2x_3-plane, i.e.

$$\begin{bmatrix} \dot{x}_2 \\ \dot{x}_3 \end{bmatrix} = \begin{bmatrix} \lambda_0 & 1 \\ 0 & \lambda_0 \end{bmatrix} \begin{bmatrix} x_2 \\ x_3 \end{bmatrix} \tag{2.108}$$

is independent of x_1. This means that each trajectory of (2.107) can be projected onto one of the trajectories of (2.108) in the x_2x_3-plane. The system (2.108) has an unstable ($\lambda_0 > 0$) improper node at $x_2 = x_3 = 0$ and so its trajectories are like those in Fig. 2.13.

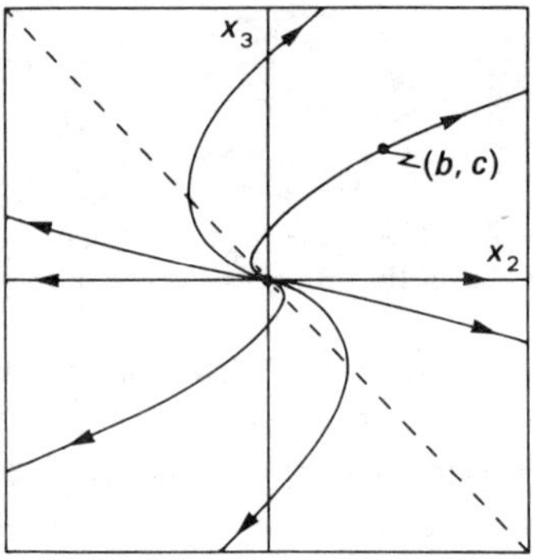

Fig. 2.13. Projections of the trajectories of system (2.107) onto the x_2x_3-plane are given by the phase portrait of (2.109), which has an unstable improper node at the origin. The line $x_3 = -\lambda_0 x_2$, where $\dot{x}_2 = 0$, is shown dashed.

Let ϕ_t denote the evolution operator for (2.107), then it follows that the trajectory $\{\phi_t\ (0, b, c) \mid t \in \mathrm{R}\}$ lies on the surface S generated by translating the trajectory,

$$\begin{bmatrix} x_2(t) \\ x_3(t) \end{bmatrix} = e^{\lambda_0 t} \begin{bmatrix} 1 & t \\ 0 & 1 \end{bmatrix} \begin{bmatrix} b \\ c \end{bmatrix}, \tag{2.109}$$

of (2.108), in the x_1-direction (see Fig. 2.14). Notice that (2.109) is the

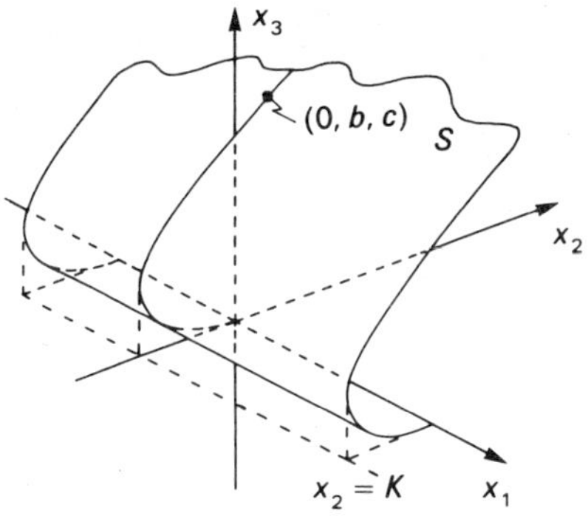

Fig. 2.14. The surface S containing the trajectory $\{\phi_t(0, b, c) \mid t \in \mathbb{R}\}$ of (2.107), obtained by translating the curve given by (2.109) in the x_1-direction.

x_2x_3-projection of

$$\phi_t(0, b, c) = e^{\lambda_0 t} \begin{bmatrix} 1 & t & \frac{1}{2}t^2 \\ 0 & 1 & t \\ 0 & 0 & 1 \end{bmatrix} \begin{bmatrix} 0 \\ b \\ c \end{bmatrix} \tag{2.110}$$

(cf. e^{Jt} in (2.106)).

Now consider the x_1x_2-projection of the required trajectory. The system equations (2.107) give $\dot{x}_1 = 0$ for $x_2 = -\lambda_0 x_1$ and $\dot{x}_2 = 0$ for $x_3 = -\lambda_0 x_2$. The latter plane intersects S in its fold line (see Fig. 2.13). The x_1x_2-projection of the fold is the line $x_2 = K$, where K is the value of x_2 for which $\dot{x}_2 = 0$ in (2.109). We can then sketch the x_1x_2-projection shown in Fig. 2.15.

Finally we can show the required trajectory on S (see Fig. 2.16). □

2.7.2 *Four-dimensional systems*

The Jordan forms $\boldsymbol{J}$ of 4×4 real matrices can be grouped as follows:

$$\text{(a)} \quad \left[\begin{array}{c|c} \boldsymbol{B} & \boldsymbol{0} \\ \hline \boldsymbol{0} & \boldsymbol{C} \end{array}\right]; \tag{2.111}$$

$$\text{(b)} \quad \left[\begin{array}{ccc|c} \lambda_0 & 1 & 0 & 0 \\ 0 & \lambda_0 & 1 & 0 \\ 0 & 0 & \lambda_0 & 0 \\ \hline 0 & 0 & 0 & \lambda_1 \end{array}\right]; \tag{2.112}$$

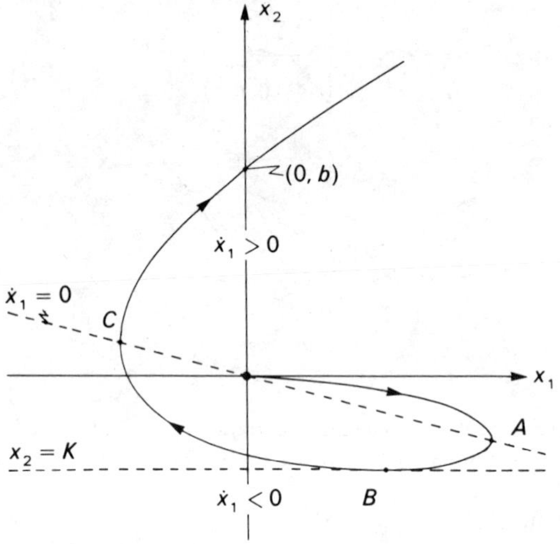

Fig. 2.15. Projection of $\{\phi_t(0, b, c) \mid t \in \mathrm{R}\}$ onto x_1, x_2-plane. A, B and C label points of the projection where $\dot{x}_1 = 0$ or $\dot{x}_2 = 0$. The turning point at $x_2 = K$ coincides with the fold in S shown in Fig. 2.14.

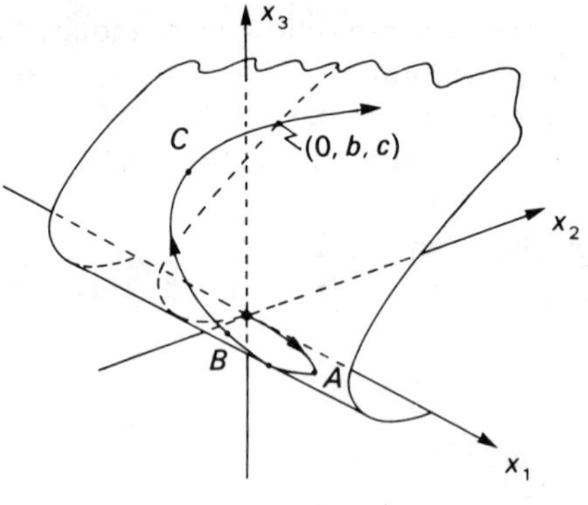

Fig. 2.16. The trajectory $\{\phi_t(0, b, c) \mid t \in \mathrm{R}\}$ of (2.107). The extreme points at A, B and C are those shown in Fig. 2.15. The curve (2.109) in the x_2, x_3-plane is shown dashed.

and

$$\text{(c)} \quad \begin{bmatrix} \lambda_0 & 1 & 0 & 0 \\ 0 & \lambda_0 & 1 & 0 \\ 0 & 0 & \lambda_0 & 1 \\ 0 & 0 & 0 & \lambda_0 \end{bmatrix} \tag{2.113}$$

In (a) the submatrices $\boldsymbol{B}$ and $\boldsymbol{C}$ are 2×2 Jordan matrices and $\boldsymbol{0}$ is the 2×2 null matrix. In (b) and (c) λ_0, λ_1 are real numbers.

The matrices in both (a) and (b) allow $\dot{\boldsymbol{y}} = \boldsymbol{J}\boldsymbol{y}$ to be split into subsystems of lower dimension. We have for (a)

$$\begin{bmatrix} \dot{x}_1 \\ \dot{x}_2 \end{bmatrix} = \boldsymbol{B}\begin{bmatrix} x_1 \\ x_2 \end{bmatrix} \quad \text{and} \quad \begin{bmatrix} \dot{x}_3 \\ \dot{x}_4 \end{bmatrix} = \boldsymbol{C}\begin{bmatrix} x_3 \\ x_4 \end{bmatrix} \tag{2.114}$$

and for (b)

$$\begin{bmatrix} \dot{x}_1 \\ \dot{x}_2 \\ \dot{x}_3 \end{bmatrix} = \begin{bmatrix} \lambda_0 & 1 & 0 \\ 0 & \lambda_0 & 1 \\ 0 & 0 & \lambda_0 \end{bmatrix} \quad \text{and} \quad \dot{x}_4 = \lambda_1 x_4. \tag{2.115}$$

The solution curves of all these subsystems have already been discussed.

The only new system we have to consider is one with a coefficient matrix of type (c). The trajectories for this system can be found by extending the method used to obtain (2.106). We find

$$\boldsymbol{x}(t) = \phi_t(\boldsymbol{x}_0) = \mathrm{e}^{\lambda_0 t}\begin{bmatrix} 1 & t & \frac{1}{2}t^2 & \frac{1}{6}t^3 \\ 0 & 1 & t & \frac{1}{2}t^2 \\ 0 & 0 & 1 & t \\ 0 & 0 & 0 & 1 \end{bmatrix}\boldsymbol{x}_0. \tag{2.116}$$

Thus it is conceptually no more difficult to find solutions for the canonical systems when $n = 4$ than it is for $n = 2$ and 3.

2.7.3 *n-Dimensional systems*

The pattern of block decomposition which we have illustrated for the Jordan forms of 3×3 and 4×4 real matrices extends to arbitrary n. This means that the corresponding canonical systems $\dot{\boldsymbol{y}} = \boldsymbol{J}\boldsymbol{y}$, with $\boldsymbol{J} = \boldsymbol{M}^{-1}\boldsymbol{A}\boldsymbol{M}$, decouple into subsystems. These subsystems may have:
dimension 1

$$\dot{y}_i = \lambda_i y_i, \qquad \lambda_i \text{ real;}$$

or dimension 2

$$\begin{bmatrix} \dot{y}_i \\ \dot{y}_{i+1} \end{bmatrix} = \begin{bmatrix} \alpha & -\beta \\ \beta & \alpha \end{bmatrix}\begin{bmatrix} y_i \\ y_{i+1} \end{bmatrix}, \qquad \beta > 0, \alpha \text{ real;}$$

or dimension j, with $2 \leqslant j \leqslant n$

$$\begin{bmatrix} \dot{y}_i \\ \cdot \\ \cdot \\ \cdot \\ \cdot \\ \cdot \\ \dot{y}_{i+j-1} \end{bmatrix} = \begin{bmatrix} \lambda & 1 & 0 & \cdot & \cdot & \cdot & 0 \\ 0 & \lambda & 1 & & & & \cdot \\ \cdot & \cdot & \cdot & & & & \cdot \\ \cdot & & 0 & \cdot & \cdot & & \cdot \\ \cdot & \cdot & & \cdot & & & \cdot \\ \cdot & \cdot & & & \cdot & \lambda & 1 \\ 0 & 0 & \cdot & \cdot & \cdot & 0 & \lambda \end{bmatrix} \begin{bmatrix} y_i \\ \cdot \\ \cdot \\ \cdot \\ \cdot \\ \cdot \\ y_{i+j-1} \end{bmatrix}, \quad \lambda \text{ real.}$$

However, as we have seen for $n = 3$ and 4, solutions can be found for all of these subsystems and we conclude any canonical system can be solved. Of course, this means that the solutions of all linear systems $\dot{\boldsymbol{x}} = \boldsymbol{A}\boldsymbol{x}$ can also be found by using $\boldsymbol{x} = \boldsymbol{M}\boldsymbol{y}$.

Exercises

Sections 2.1 *and* 2.2

1. Indicate the effect of each of the following linear transformations of the x_1x_2-plane by shading the image of the square $S = \{(x_1, x_2) | 0 \leqslant x_1, x_2 \leqslant 1\}$:

(a) $\begin{bmatrix} 1 & 0 \\ 0 & 2 \end{bmatrix}$; (b) $\begin{bmatrix} -1 & 2 \\ -2 & -1 \end{bmatrix}$; (c) $\begin{bmatrix} -7 & 3 \\ -8 & 3 \end{bmatrix}$; (d) $\begin{bmatrix} 1 & -2 \\ 1 & 1 \end{bmatrix}$.

Use the images of the square to help to sketch the images of the following sets:
(a) the circle $x_1^2 + x_2^2 = 1$; (b) the curve $x_1x_2 = 1, \quad x_1, x_2 > 0$.

2. Prove that the relation $\sim$ on pairs $\boldsymbol{A}$, $\boldsymbol{B}$ or real $n \times n$ matrices given by

$$\boldsymbol{A} \sim \boldsymbol{B} \Leftrightarrow \boldsymbol{B} = \boldsymbol{M}^{-1}\boldsymbol{A}\boldsymbol{M},$$

where $\boldsymbol{M}$ is a non-singular matrix, is an equivalence relation. Furthermore, show that all the matrices in any given equivalence class have the same eigenvalues. Group the following systems under this equivalence on their coefficient matrices:

(a) $\dot{\boldsymbol{x}} = \begin{bmatrix} 2 & 0 \\ 0 & 2 \end{bmatrix}\boldsymbol{x}$; (b) $\dot{\boldsymbol{x}} = \begin{bmatrix} 2 & 1 \\ 0 & 2 \end{bmatrix}\boldsymbol{x}$;

(c) $\dot{\boldsymbol{x}} = \begin{bmatrix} 7 & -2 \\ 15 & -6 \end{bmatrix}\boldsymbol{x}$; (d) $\dot{\boldsymbol{x}} = \begin{bmatrix} -4 & 2 \\ -4 & 5 \end{bmatrix}\boldsymbol{x}$;

(e) $\dot{\boldsymbol{x}} = \begin{bmatrix} 2 & 2 \\ 3 & 1 \end{bmatrix}\boldsymbol{x}$; (f) $\dot{\boldsymbol{x}} = \begin{bmatrix} 2 & 2 \\ 0 & 2 \end{bmatrix}\boldsymbol{x}$.

3. Show that each of the following systems can be transformed into a canonical system by the given change of variable:

(a) $\dot{x}_1 = 4x_1 + x_2, \quad \dot{x}_2 = -x_1 + 2x_2,$
$y_1 = 2x_1 + x_2, \quad y_2 = x_1 + x_2;$
(b) $\dot{x}_1 = 12x_1 + 4x_2, \quad \dot{x}_2 = -26x_1 - 8x_2,$
$y_1 = 2x_1 + x_2, \quad y_2 = 3x_1 + x_2;$
(c) $\dot{x}_1 = 10x_1 + 2x_2, \quad \dot{x}_2 = -28x_1 - 5x_2,$
$y_1 = 7x_1 + 2x_2, \quad y_2 = 4x_1 + x_2.$

Write down the coefficient matrices $\boldsymbol{A}$ of each of the above systems, the Jordan forms $\boldsymbol{J}$ and the change of variable matrices $\boldsymbol{M}$. Check that $\boldsymbol{J} = \boldsymbol{M}^{-1}\boldsymbol{A}\boldsymbol{M}$ is satisfied.

4. For each matrix $\boldsymbol{A}$ given below

$$\begin{bmatrix} 0 & 1 \\ -2 & 3 \end{bmatrix}, \quad \begin{bmatrix} 41 & -29 \\ 58 & -41 \end{bmatrix}, \quad \begin{bmatrix} 9 & 4 \\ -9 & -3 \end{bmatrix}$$

find its Jordan form $\boldsymbol{J}$ and a matrix $\boldsymbol{M}$ that satisfies

$$\boldsymbol{J} = \boldsymbol{M}^{-1}\boldsymbol{A}\boldsymbol{M}.$$

5. What is the effect on the system

$$\begin{aligned} \dot{x}_1 &= -7x_1 - 4x_2 - 6x_3 \\ \dot{x}_2 &= -3x_1 - 2x_2 - 3x_3 \\ \dot{x}_3 &= 3x_1 + 2x_2 + 2x_3 \end{aligned}$$

of the change of variables $y_1 = x_1 + 2x_3$, $y_2 = x_2$, $y_3 = x_1 + x_3$? How does the introduction of these new variables help to investigate the solution curves of the system?

6. Prove that no two different 2×2 Jordan matrices are similar. Hence prove that there is one and only one Jordan matrix similar to any given 2×2 real matrix.

7. Find the characteristic equations of the matrices

$$\begin{bmatrix} 2 & 1 \\ 0 & 2 \end{bmatrix}, \quad \begin{bmatrix} 3 & 1 \\ -2 & 7 \end{bmatrix}, \quad \begin{bmatrix} 8 & -4 \\ 6 & -3 \end{bmatrix},$$

and illustrate the Cayley–Hamilton theorem for each matrix.

Section 2.3

8. Find the solution curves $\boldsymbol{x}(t)$ satisfying $\dot{\boldsymbol{x}} = \boldsymbol{A}\boldsymbol{x}$ subject to $\boldsymbol{x}(0) = \boldsymbol{x}_0$, where the matrix $\boldsymbol{A}$ is taken to be each of the matrices in Exercise 2.4.

9. Sketch phase portraits for the linear system

$$\begin{bmatrix} \dot{x}_1 \\ \dot{x}_2 \end{bmatrix} = \boldsymbol{A} \begin{bmatrix} x_1 \\ x_2 \end{bmatrix}$$

when $\boldsymbol{A}$ is given by:

(a) $\begin{bmatrix} 1 & 0 \\ 0 & 2 \end{bmatrix}$; (b) $\begin{bmatrix} -1 & 0 \\ 0 & 2 \end{bmatrix}$; (c) $\begin{bmatrix} 3 & 1 \\ -1 & 3 \end{bmatrix}$;

(d) $\begin{bmatrix} 2 & 1 \\ 0 & 2 \end{bmatrix}$; (e) $\begin{bmatrix} 1 & 0 \\ 0 & \frac{1}{2} \end{bmatrix}$; (f) $\begin{bmatrix} -3 & 0 \\ 0 & -3 \end{bmatrix}$;

(g) $\begin{bmatrix} 0 & 2 \\ -2 & 0 \end{bmatrix}$; (h) $\begin{bmatrix} 3 & 0 \\ 0 & -1 \end{bmatrix}$; (i) $\begin{bmatrix} 3 & 0 \\ 0 & 0 \end{bmatrix}$;

(j) $\begin{bmatrix} 0 & -2 \\ 2 & 0 \end{bmatrix}$.

10. Indicate the effect of the linear transformation

$$\begin{bmatrix} y_1 \\ y_2 \end{bmatrix} = \begin{bmatrix} 2 & 1 \\ 1 & 1 \end{bmatrix} \begin{bmatrix} x_1 \\ x_2 \end{bmatrix}$$

on each of the systems in Exercise 2.9 by sketching each phase portrait in the $y_1 y_2$-plane.

11. Find the 2×2 matrix $\boldsymbol{A}$ such that the system

$$\dot{\boldsymbol{x}} = \boldsymbol{A}\boldsymbol{x}$$

has a solution curve

$$\boldsymbol{x}(t) = \begin{bmatrix} e^{-t}(\cos t + 2\sin t) \\ e^{-t}\cos t \end{bmatrix}.$$

12. Let

$$\dot{\boldsymbol{x}} = \begin{bmatrix} -1 & -1 \\ 2 & -4 \end{bmatrix}\boldsymbol{x}$$

and $y = x_1 + 3x_2$. For this situation the system is said to be 'observable' if by knowing $y(t)$ the solution curve $\boldsymbol{x}(t)$ can be derived. Prove that the system is observable when $y(t) = 4e^{-2t}$ by finding $\boldsymbol{x}(0)$.

Section 2.4

13. Locate each of the following linear systems in the Tr–Det plane and hence state their phase portrait type:
(a) $\dot{x}_1 = 2x_1 + x_2, \quad \dot{x}_2 = x_1 + 2x_2;$
(b) $\dot{x}_1 = 2x_1 + x_2, \quad \dot{x}_2 = x_1 - 3x_2;$
(c) $\dot{x}_1 = x_1 - 4x_2, \quad \dot{x}_2 = 2x_1 - x_2;$
(d) $\dot{x}_1 = 2x_2, \quad \dot{x}_2 = -3x_1 - x_2,$
(e) $\dot{x}_1 = -x_1 + 8x_2, \quad \dot{x}_2 = -2x_1 + 7x_2.$

14. Express the system

$$\dot{x}_1 = -7x_1 + 3x_2, \qquad \dot{x}_2 = -8x_1 + 3x_2$$

in coordinates y_1, y_2 relative to the basis $\{(1, 2), (3, 4)\}$. Hence sketch the phase portrait of the system in the x_1x_2-plane.

15. Let $\boldsymbol{x}(t)$ be a trajectory of the linear system $\dot{\boldsymbol{x}} = \boldsymbol{A}\boldsymbol{x}$ and suppose $\lim_{t\to\infty}\boldsymbol{x}(t) = \boldsymbol{0}$. Use the continuity of the linear transformation $\boldsymbol{y} = \boldsymbol{N}\boldsymbol{x}$, where $\boldsymbol{N}$ is a non-singular matrix, to show that the trajectory $\boldsymbol{y}(t) = \boldsymbol{N}\boldsymbol{x}(t)$ of the system

$$\dot{\boldsymbol{y}} = \boldsymbol{N}\boldsymbol{A}\boldsymbol{N}^{-1}\boldsymbol{y}$$

also satisfies the property

$$\lim_{t\to\infty}\boldsymbol{y}(t) = \boldsymbol{0}.$$

Hence prove that if $\dot{\boldsymbol{x}} = \boldsymbol{A}\boldsymbol{x}$ has a fixed point which is a stable node, improper node or focus then so has the system $\dot{\boldsymbol{y}} = \boldsymbol{N}\boldsymbol{A}\boldsymbol{N}^{-1}\boldsymbol{y}$.

16. Find a corresponding result to that of Exercise 2.15 when a trajectory $\boldsymbol{x}(t)$ of the system $\dot{\boldsymbol{x}} = \boldsymbol{A}\boldsymbol{x}$ satisfies $\lim_{t \to -\infty} \boldsymbol{x}(t) = \boldsymbol{0}$. Use both results to prove that if the system $\dot{\boldsymbol{x}} = \boldsymbol{A}\boldsymbol{x}$ has a saddle point at the origin then so does the system $\dot{\boldsymbol{y}} = \boldsymbol{N}\boldsymbol{A}\boldsymbol{N}^{-1}\boldsymbol{y}$.

17. Show that there is a linear mapping of the family of curves $x_2 = Cx_1^{\mu}$ onto the family of curves $y_2 = C'y_1^{\mu'}$ (μ, μ' constant) if and only if $\mu = \mu'$ or $\mu = 1/\mu'$. Hence show that there is a linear mapping of the trajectories of

$$\dot{x}_1 = \lambda_1 x_1, \qquad \dot{x}_2 = \lambda_2 x_2, \qquad \lambda_1 \lambda_2 \neq 0,$$

onto the phase portrait of the system

$$\dot{y}_1 = \nu_1 y_1, \qquad \dot{y}_2 = \nu_2 y_2$$

if and only if $\lambda_1/\lambda_2 = \nu_1/\nu_2$ or $\lambda_1/\lambda_2 = \nu_2/\nu_1$.

18. Show that there is a real positive number k such that the continuous bijection of the plane

$$y_1 = x_1$$

$$y_2 = \begin{cases} x_2^k, & x_2 \geqslant 0 \\ -(-x_2)^k, & x_2 < 0 \end{cases}$$

maps the trajectories of the node $\dot{x}_1 = \lambda_1 x_1$, $\dot{x}_2 = \lambda_2 x_2$ onto the trajectories of the star node $\dot{y}_1 = \varepsilon y_1$, $\dot{y}_2 = \varepsilon y_2$, $\varepsilon = +1$ or -1, preserving orientation.

19. Show that the results developed in Exercise 2.18 can be used to find a mapping which maps trajectories of any saddle point onto the trajectories of the saddle $\dot{x}_1 = x_1$, $\dot{x}_2 = -x_2$.

20. Sketch the phase portrait of the system

$$\dot{x}_1 = 2x_1 + x_2, \qquad \dot{x}_2 = 3x_2$$

and the phase portraits obtained by

(a) reflection in the x_1-axis;
(b) a half turn in the x_1x_2-plane;
(c) an anticlockwise rotation of $\pi/2$;
(d) interchanging the axes x_1 and x_2.

State transforming matrices for each of the cases above and find the transformed system equations. Check that each of the phase portraits you have obtained corresponds to the appropriate transformed system.

Section 2.5

21. Calculate e^{At} for the following matrices A:

(a) $\begin{bmatrix} 2 & 0 \\ 0 & 3 \end{bmatrix}$, (b) $\begin{bmatrix} 1 & 2 \\ 0 & 2 \end{bmatrix}$, (c) $\begin{bmatrix} 2 & 4 \\ 3 & 3 \end{bmatrix}$,

(d) $\begin{bmatrix} -2 & 2 \\ -4 & -2 \end{bmatrix}$, (e) $\begin{bmatrix} -4 & 1 \\ -1 & -2 \end{bmatrix}$,

22. Calculate e^{At} when A is equal to:

(a) $\begin{bmatrix} 2 & -7 \\ 3 & -8 \end{bmatrix}$; (b) $\begin{bmatrix} 2 & 4 \\ -2 & 6 \end{bmatrix}$.

Use the results to calculate e^{At} when A equals:

(c) $\begin{bmatrix} 2 & -7 & 0 \\ 3 & -8 & 0 \\ 0 & 0 & 1 \end{bmatrix}$; (d) $\begin{bmatrix} 2 & -7 & 0 & 0 \\ 3 & -8 & 0 & 0 \\ 0 & 0 & 2 & 4 \\ 0 & 0 & -2 & 6 \end{bmatrix}$.

23. Re-order the variables of the system

$$\dot{\boldsymbol{x}} = \begin{bmatrix} 2 & 0 & 1 & 0 \\ 0 & -5 & 0 & 9 \\ 0 & 0 & 2 & 0 \\ 0 & -4 & 0 & 8 \end{bmatrix} \boldsymbol{x}$$

so that the coefficient matrix takes the same form as that in Exercise 2.22(d). Hence, or otherwise, find the trajectory passing through (2, 1, 0, 1) at $t = 1$.

24. Prove that if $\boldsymbol{P}$ and $\boldsymbol{Q}$ are $n \times n$ commuting matrices then $\boldsymbol{P}^r\boldsymbol{Q}^s = \boldsymbol{Q}^s\boldsymbol{P}^r$ for r, s non-negative integers.

Hence show by induction on n, or otherwise, that

$$\sum_{k=0}^{n} \frac{(\boldsymbol{P}+\boldsymbol{Q})^k}{k!} = \left(\sum_{k=0}^{n} \frac{\boldsymbol{P}^k}{k!}\right)\left(\sum_{k=0}^{n} \frac{\boldsymbol{Q}^k}{k!}\right)$$

and deduce

$$\mathrm{e}^{\boldsymbol{P}+\boldsymbol{Q}} = \mathrm{e}^{\boldsymbol{P}}\mathrm{e}^{\boldsymbol{Q}}.$$

25. Use (2.63) to evaluate $\mathrm{e}^{\boldsymbol{A}t}$ when:

(a) $\boldsymbol{A} = \begin{bmatrix} \lambda_0 & 1 \\ 0 & \lambda_0 \end{bmatrix}$ by writing $\boldsymbol{A} = \lambda_0 \boldsymbol{I} + \boldsymbol{C}$ where $\boldsymbol{C} = \begin{bmatrix} 0 & 1 \\ 0 & 0 \end{bmatrix}$;

(b) $\boldsymbol{A} = \begin{bmatrix} \alpha & -\beta \\ \beta & \alpha \end{bmatrix}$ by writing $\boldsymbol{A} = \alpha \boldsymbol{I} + \beta \boldsymbol{D}$, where

$$\boldsymbol{D} = \begin{bmatrix} 0 & -1 \\ 1 & 0 \end{bmatrix}.$$

26. Write down the characteristic equation of a 2×2 matrix $\boldsymbol{A}$. Use the Cayley–Hamilton theorem to obtain an expression for $\boldsymbol{A}^2$ in terms of $\boldsymbol{A}$ and $\boldsymbol{I}$. Use this result to prove that if λ_1, λ_2 ($\lambda_1 \neq \lambda_2$) are the eigenvalues of $\boldsymbol{A}$, then

$$\left\{\frac{\boldsymbol{A} - \lambda_1 \boldsymbol{I}}{\lambda_2 - \lambda_1}\right\}\left\{\frac{\boldsymbol{A} - \lambda_2 \boldsymbol{I}}{\lambda_1 - \lambda_2}\right\} = \boldsymbol{0},$$

and

$$\left\{\frac{\boldsymbol{A} - \lambda_1 \boldsymbol{I}}{\lambda_2 - \lambda_1}\right\}^2 = \left\{\frac{\boldsymbol{A} - \lambda_1 \boldsymbol{I}}{\lambda_2 - \lambda_1}\right\}$$

(c.f. Section 2.5).

27. Let $\boldsymbol{A}$ be a 3×3 real matrix with eigenvalues $\lambda_1 = \lambda_2 = \lambda_3$ $(= \lambda_0)$. Show that

$$\mathrm{e}^{\boldsymbol{A}t} = \mathrm{e}^{\lambda_0 t}(\boldsymbol{I} + t\boldsymbol{Q} + \tfrac{1}{2}t^2\boldsymbol{Q}^2)$$

where $\boldsymbol{Q} = \boldsymbol{A} - \lambda_0 \boldsymbol{I}$. Generalize this result to $n \times n$ matrices.

Section 2.6

28. Find, where possible, a change of variables which converts each affine system into an equivalent linear system.
(a) $\dot{x}_1 = x_1 + x_2 + 2, \quad \dot{x}_2 = x_1 + 2x_2 + 3;$
(b) $\dot{x}_1 = x_2 + 1, \quad \dot{x}_2 = 3;$
(c) $\dot{x}_1 = 2x_1 - 3x_2 + 1, \quad \dot{x}_2 = 6x_1 - 9x_2;$
(d) $\dot{x}_1 = 2x_1 - x_2 + 1, \quad \dot{x}_2 = 6x_1 + 3x_2;$
(e) $\dot{x}_1 = 2x_1 + x_2 + 1, \quad \dot{x}_2 = x_1 + x_3, \quad \dot{x}_3 = x_2 + x_3 + 2;$
(f) $\dot{x}_1 = x_1 + x_2 + x_3 + 1, \quad \dot{x}_2 = -x_2, \quad \dot{x}_3 = x_1 + x_3 + 1.$
For those affine systems equivalent to a linear system, state their algebraic type.

29. Find the solution curve $(x_1(t), x_2(t))$ which satisfies

$$\dot{x}_1 = x_1 + x_2 + 1, \qquad \dot{x}_2 = x_1 + x_2$$

subject to the initial condition $x_1(0) = a$, $x_2(0) = b$.

30. When the non-singular change of variables $\boldsymbol{x} = \boldsymbol{My}$, $\boldsymbol{x}, \boldsymbol{y} \in \mathrm{R}^n$, is applied to the affine system $\dot{\boldsymbol{x}} = \boldsymbol{Ax} + \boldsymbol{h}$, what is the transformed system? Hence show, when $n = 2$, that every affine system can be changed to an affine system with a Jordan coefficient matrix. In the case when $\boldsymbol{A}$ has two real distinct eigenvalues show that the system $\dot{\boldsymbol{x}} = \boldsymbol{Ax} + \boldsymbol{h}(t)$ can be decoupled.

Is it possible to change an affine system to a linear system by a linear transformation?

31. Sketch phase portraits of the affine systems
(a) $\dot{x}_1 = 2x_2 + 1, \quad \dot{x}_2 = -x_1 + 1$ (b) $\dot{x}_1 = x_1 + 2, \quad \dot{x}_2 = 3.$

Section 2.7

32. Show that the system

$$\begin{aligned} \dot{x}_1 &= -3x_1 + 10x_2 \\ \dot{x}_2 &= -2x_1 + 5x_2 \\ \dot{x}_3 &= -2x_1 + 2x_2 + 3x_3 \end{aligned}$$

can be transformed into

$$\begin{aligned} \dot{y}_1 &= ay_1 - by_2 \\ \dot{y}_2 &= by_1 + ay_2 \\ \dot{y}_3 &= cy_3 \end{aligned}$$

where a, b and c are constants. What are their values? Find a matrix $\boldsymbol{M}$ of the form

$$\begin{bmatrix} m_{11} & m_{12} & 0 \\ 0 & m_{22} & 0 \\ 0 & m_{32} & m_{33} \end{bmatrix}$$

such that

$$\boldsymbol{x} = \boldsymbol{M}\boldsymbol{y}.$$

What form does the phase portrait take when projected onto the y_1y_2-plane and the y_3-axis respectively? Make a sketch of some trajectories of the transformed system.

33. Find a solution curve of the system

$$\begin{aligned} \dot{x}_1 &= x_1 + 2x_2 + x_3 \\ \dot{x}_2 &= x_2 + 2x_3 \\ \dot{x}_3 &= 2x_3 \end{aligned}$$

which satisfies $x_1(0) = x_2(0) = 0$ and $x_3(0) = 1$.

34. Find the form of the solution curves of the systems $\dot{\boldsymbol{x}} = \boldsymbol{A}\boldsymbol{x}$, $\boldsymbol{x} \in \mathsf{R}^4$, where $\boldsymbol{A}$ equals:

$$\text{(a)} \begin{bmatrix} \lambda & 1 & 0 & 0 \\ 0 & \lambda & 1 & 0 \\ 0 & 0 & \lambda & 1 \\ 0 & 0 & 0 & \lambda \end{bmatrix}; \quad \text{(b)} \begin{bmatrix} \alpha & -\beta & 0 & 0 \\ \beta & \alpha & 0 & 0 \\ 0 & 0 & \lambda & 1 \\ 0 & 0 & 0 & \lambda \end{bmatrix}; \quad \text{(c)} \begin{bmatrix} \alpha & -\beta & 0 & 0 \\ \beta & \alpha & 0 & 0 \\ 0 & 0 & \gamma & -\delta \\ 0 & 0 & \delta & \gamma \end{bmatrix}.$$

35. Consider the six-dimensional system

$$\begin{bmatrix} \dot{x}_1 \\ \dot{x}_2 \\ \dot{x}_3 \\ \dot{x}_4 \\ \dot{x}_5 \\ \dot{x}_6 \end{bmatrix} = \begin{bmatrix} 2 & 0 & 0 & 0 & 0 & 0 \\ 0 & 0 & -1 & 0 & 0 & 0 \\ 0 & 1 & 0 & 0 & 0 & 0 \\ 0 & 0 & 0 & 1 & 0 & 0 \\ 0 & 0 & 0 & 0 & 1 & 1 \\ 0 & 0 & 0 & 0 & 0 & 1 \end{bmatrix} \begin{bmatrix} x_1 \\ x_2 \\ x_3 \\ x_4 \\ x_5 \\ x_6 \end{bmatrix}$$

Divide the system into subsystems and describe the phase portrait behaviour of each subsystem.

CHAPTER THREE

Non-linear systems in the plane

In this chapter we consider the phase portraits of systems $\dot{\mathbf{x}} = \mathbf{X}(\mathbf{x})$, $\mathbf{x} \in S \subseteq \mathsf{R}^2$, where $\mathbf{X}$ is a continuously differentiable, non-linear function. In contrast to Sections 1.2 and 2.3 these phase portraits are *not* always determined by the nature of the fixed points of the system.

3.1 Local and global behaviour

Definition 3.1.1. A *neighbourhood*, N, of a point $\mathbf{x}_0 \in \mathsf{R}^2$ is a subset of R^2 containing a disc $\{\mathbf{x} \,|\, |\mathbf{x} - \mathbf{x}_0| < r\}$ for some $r > 0$.

Definition 3.1.2. The part of the phase portrait of a system that occurs in a neighbourhood N of $\mathbf{x}_0$ is called the *restriction* of the phase portrait to N.

These definitions are illustrated in Fig. 3.1 using a simple linear system.

When analysing non-linear systems, we often consider a restriction of the complete or *global* phase portrait to a neighbourhood of $\mathbf{x}_0$ that is as small as we please (see Section 3.3). Such a restriction will be referred to as the *local phase portrait at* $\mathbf{x}_0$.

Consider the restriction of a simple linear system to a neighbourhood N of the origin. There is a neighbourhood $N' \subseteq N$ such that the restriction of this phase portrait to N' is qualitatively equivalent to the global phase portrait of the simple linear system itself. That is, there is a continuous bijection between N' and R^2 which

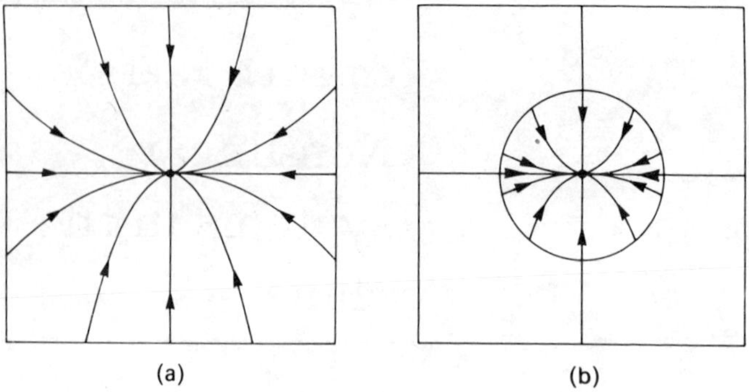

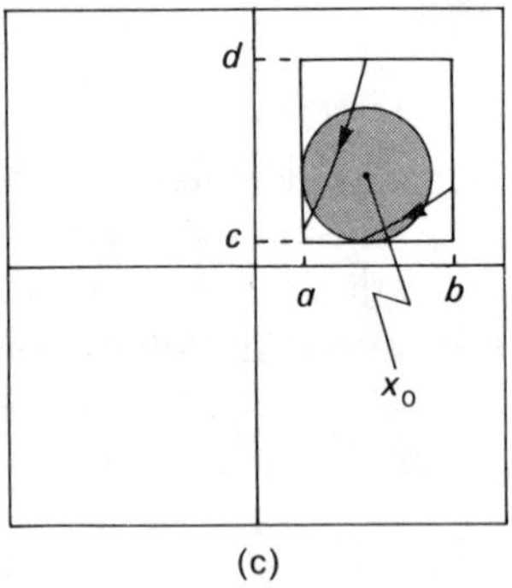

Fig. 3.1. (a) Phase portrait for $\dot{x}_1 = -x_1$, $\dot{x}_2 = -2x_2$. (b) Restriction of (a) to $N = \{\mathbf{x} \,|\, |\mathbf{x}| < a\}$, $a > 0$, of the fixed point at (0, 0). (c) Restriction of (a) to $N = \{\mathbf{x} \,|\, a < x_1 < b,\ \ c < x_2 < d\}$ $a, b, c, d > 0$, of point $\mathbf{x}_0$ where $\dot{\mathbf{x}} \neq \mathbf{0}$. A disc radius $r > 0$ centred on $\mathbf{x}_0$ is shown shaded.

maps the phase portrait restricted to N' onto the complete phase portrait. This result is illustrated for a centre in Fig. 3.2. The neighbourhood N is taken to be $\{(x_1, x_2) \,|\, a < x_1 < b,\ c < x_2 < d;\ a, c < 0; b, d < 0\}$. Let N' be the set of (x_1, x_2) lying inside the critical trajectory T shown in Fig. 3.2(b). Every trajectory in the global phase portrait (Fig. 3.2(a)) has a counterpart in its restriction to N' and *vice versa.* If we consider the restriction of the stable node shown in Fig. 3.1(b), then again the result holds but in this case with $N' = N$.

This qualitative equivalence of the phase portrait and its restrictions is what we really mean by saying that the phase portrait of a simple linear system is determined by the 'nature' of its fixed point. In other words, the *local* phase portrait at the origin is qualitatively equivalent to the *global* phase portrait of the system.

Non-linear systems can have more than one fixed point and we can often obtain the local phase portraits at all of them. However, as Fig. 3.3 shows, local phase portraits do *not* always determine the

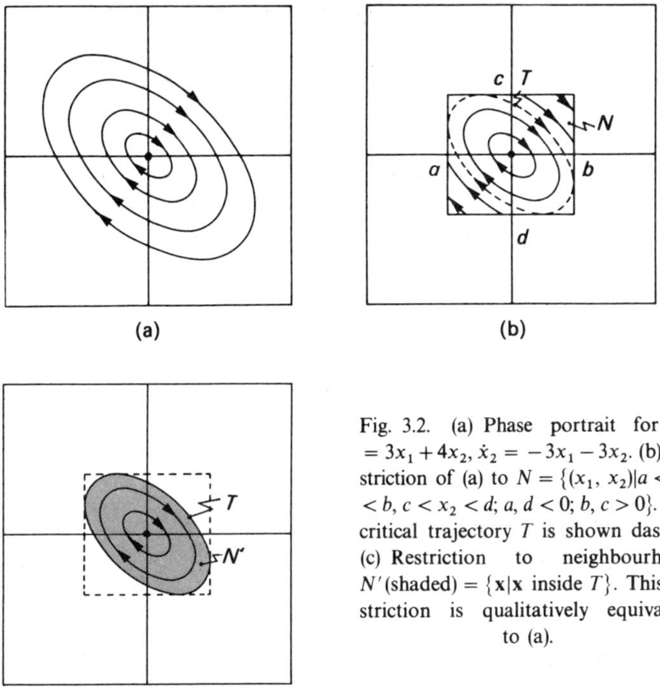

(a) (b) (c)

Fig. 3.2. (a) Phase portrait for $\dot{x}_1 = 3x_1 + 4x_2$, $\dot{x}_2 = -3x_1 - 3x_2$. (b) Restriction of (a) to $N = \{(x_1, x_2) | a < x_1 < b, c < x_2 < d; a, d < 0; b, c > 0\}$. The critical trajectory T is shown dashed. (c) Restriction to neighbourhood N'(shaded) $= \{\mathbf{x} | \mathbf{x}$ inside $T\}$. This restriction is qualitatively equivalent to (a).

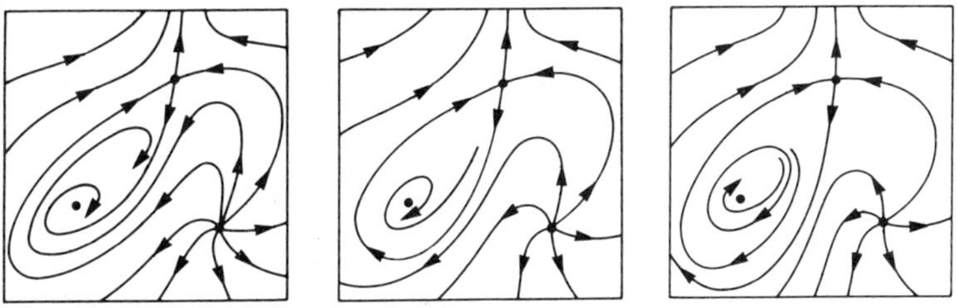

(a) (b) (c)

Fig. 3.3. Qualitatively different global phase portraits consistent with given local behaviour at three fixed points.

global phase portrait. The figure shows three qualitatively *different global* phase portraits, each containing three fixed points. The *local* phase portraits at the fixed points are the *same* in all three diagrams.

These phase portraits are realized by the non-linear system

$$\begin{aligned}\dot{x}_1 &= -\alpha x_2 + x_1(1 - x_1^2 - x_2^2) - x_2(x_1^2 + x_2^2)\\ \dot{x}_2 &= \alpha x_1 + x_2(1 - x_1^2 - x_2^2) + x_1(x_1^2 + x_2^2) + \beta\end{aligned} \tag{3.1}$$

for certain choices of the pair (α, β).

Figure 3.3(c) also illustrates another global feature of non-linear phase portraits which is not revealed by a study of fixed points. The isolated closed orbit around one of the fixed points is called a *limit cycle*. The detection of limit cycles requires a global approach (see Section 3.9).

The treatment of non-linear systems, therefore, involves techniques relating to both local and global behaviour. The former are discussed in Sections 3.2–3.6.1 inclusive, while the latter are dealt with in Sections 3.6.2–3.9.

3.2 Linearization at a fixed point

We begin by examining non-linear systems with a fixed point at the origin of their phase plane.

Definition 3.2.1. Suppose the system $\dot{\mathbf{y}} = \mathbf{Y}(\mathbf{y})$ can be written in the form

$$\begin{aligned}\dot{y}_1 &= ay_1 + by_2 + g_1(y_1, y_2)\\ \dot{y}_2 &= cy_1 + dy_2 + g_2(y_1, y_2),\end{aligned} \tag{3.2}$$

where $[g_i(y_1, y_2)/r] \to 0$ as $r = (y_1^2 + y_2^2)^{1/2} \to 0$. The linear system

$$\dot{y}_1 = ay_1 + by_2, \qquad \dot{y}_2 = cy_1 + dy_2 \tag{3.3}$$

is said to be the *linearization* (or *linearized system*) of (3.2) at the origin. The components of the linear vector field in (3.3) are said to form the *linear part* of $\mathbf{Y}$.

Example 3.2.1. Find the linearizations of the following systems:

(a) $\dot{y}_1 = y_1 + y_1^2 + y_1 y_2^2, \quad \dot{y}_2 = y_2 + y_2^{3/2}$;
(b) $\dot{y}_1 = y_1^3, \quad \dot{y}_2 = y_2 + y_2 \sin y_1$;
(c) $\dot{y}_1 = y_1^2 e^{y_2}, \quad \dot{y}_2 = y_2(e^{y_1} - 1)$.

Solution. For each system, we tabulate the real numbers a, b, c, d and the functions g_1 and g_2 below. The functions g_i $(i = 1, 2)$ are given in polar coordinates so that the requirement

$$\mathrm{Lim}_{r\to 0}\,[g_i(y_1, y_2)/r] = 0, \qquad i = 1, 2, \tag{3.4}$$

can be checked.

System	a	b	c	d	g_1	g_2
(a)	1	0	0	1	$r^2(\cos^2\theta + r\cos\theta\sin^2\theta)$	$r^{3/2}\sin^{3/2}\theta$
(b)	0	0	0	1	$r^3\cos^3\theta$	$r\sin\theta\sin(r\cos\theta)$
(c)	0	0	0	0	$r^2\cos^2\theta\, e^{r\sin\theta}$	$r\sin\theta(r\cos\theta + + \frac{1}{2}r^2\cos^2\theta + \ldots)$

Hence, the linearizations are:

(a) $\dot{y}_1 = y_1, \quad \dot{y}_2 = y_2;$
(b) $\dot{y}_1 = 0, \quad \dot{y}_2 = y_2;$
(c) $\dot{y}_1 = 0, \quad \dot{y}_2 = 0.$ □

Definition 3.2.1 can be applied at fixed points which do not occur at the origin by introducing *local coordinates.* Suppose (ξ, η) is a fixed point of the non-linear system $\dot{\mathbf{x}} = \mathbf{X}(\mathbf{x})$, $\mathbf{x} = (x_1, x_2)$. The variables

$$y_1 = x_1 - \xi, \qquad y_2 = x_2 - \eta \tag{3.5}$$

are a set of Cartesian coordinates for the phase plane with their origin at $(x_1, x_2) = (\xi, \eta)$. They are said to be local coordinates at (ξ, η). It follows that

$$\dot{y}_i = \dot{x}_i = X_i(y_1 + \xi, y_2 + \eta), \qquad i = 1, 2, \tag{3.6}$$

where X_1, X_2 are the component functions of $\mathbf{X}$. If we define

$$Y_i(y_1, y_2) = X_i(y_1 + \xi, y_2 + \eta), \tag{3.7}$$

(3.6) becomes

$$\dot{y}_i = Y_i(y_1, y_2), \qquad i = 1, 2 \quad \text{or} \quad \dot{\mathbf{y}} = \mathbf{Y}(\mathbf{y}). \tag{3.8}$$

The system in (3.8) has the fixed point of interest at the origin of its phase plane and Definition 3.2.1 can be applied.

Example 3.2.2. Show that the system

$$\dot{x}_1 = e^{x_1 + x_2} - x_2, \qquad \dot{x}_2 = -x_1 + x_1 x_2 \tag{3.9}$$

has only one fixed point. Find the linearization of (3.9) at this point.

Solution. The fixed points of the system satisfy

$$e^{x_1+x_2} - x_2 = 0 \tag{3.10}$$

and

$$x_1(x_2 - 1) = 0. \tag{3.11}$$

Equation (3.11) is satisfied only by $x_1 = 0$ or $x_2 = 1$. If $x_1 = 0$, (3.10) becomes $e^{x_2} = x_2$ which has no real solution ($e^{x_2} > x_2$, for all real x_2) and there is no fixed point with $x_1 = 0$. If $x_2 = 1$, (3.10) gives $e^{x_1+1} = 1$ which has only one real solution, $x_1 = -1$. Thus $(x_1, x_2) = (-1, 1)$ is the only fixed point of (3.9).

To find the linearized system at $(-1, 1)$, introduce local coordinates $y_1 = x_1 + 1$ and $y_2 = x_2 - 1$. We find

$$\dot{y}_1 = e^{y_1+y_2} - y_2 - 1, \qquad \dot{y}_2 = -y_2 + y_1 y_2. \tag{3.12}$$

This can be written in the form (3.2), by using the power series expansion of $e^{y_1+y_2}$;

$$\begin{aligned} \dot{y}_1 &= y_1 + \left\{ \frac{(y_1+y_2)^2}{2!} + \frac{(y_1+y_2)^3}{3!} + \dots \right\} \\ \dot{y}_2 &= -y_2 + y_1 y_2. \end{aligned} \tag{3.13}$$

Finally, we recognize the linearization as

$$\dot{y}_1 = y_1, \qquad \dot{y}_2 = -y_2. \tag{3.14}$$ □

Example 3.2.2 suggests a systematic way of obtaining linearizations by utilizing Taylor expansions. If the component functions $X_i(x_1, x_2)$ $(i = 1, 2)$ are continuously differentiable in some neighbourhood of the point (ξ, η) then for each i

$$\begin{aligned} X_i(x_1, x_2) = X_i(\xi, \eta) &+ (x_1 - \xi)\frac{\partial X_i}{\partial x_1}(\xi, \eta) \\ &+ (x_2 - \eta)\frac{\partial X_i}{\partial x_2}(\xi, \eta) + R_i(x_1, x_2). \end{aligned} \tag{3.15}$$

The remainder functions $R_i(x_1, x_2)$ satisfy

$$\mathrm{Lim}_{r \to 0}\,[R_i(x_1, x_2)/r] = 0, \tag{3.16}$$

where $r = \{(x-\xi)^2 + (y-\eta)^2\}^{1/2}$. If (ξ, η) is a fixed point of $\dot{\mathbf{x}} = \mathbf{X}(\mathbf{x})$, then $X_i(\xi, \eta) = 0$ $(i = 1, 2)$ and on introducing local coordinates (3.5), we obtain

$$\begin{aligned} \dot{y}_1 &= y_1 \frac{\partial X_1}{\partial x_1}(\xi, \eta) + y_2 \frac{\partial X_1}{\partial x_2}(\xi, \eta) + R_1(y_1 + \xi, y_2 + \eta) \\ \dot{y}_2 &= y_1 \frac{\partial X_2}{\partial x_1}(\xi, \eta) + y_2 \frac{\partial X_2}{\partial x_2}(\xi, \eta) + R_2(y_1 + \xi, y_2 + \eta). \end{aligned} \tag{3.17}$$

Equation (3.16) ensures that (3.17) is in the form (3.2) with $g_i(y_1, y_2) \equiv R_i(y_1 + \xi, y_2 + \eta)$ $(i = 1, 2)$ and the linearization at (ξ, η) is given by

$$a = \frac{\partial X_1}{\partial x_1}, \qquad b = \frac{\partial X_1}{\partial x_2}, \qquad c = \frac{\partial X_2}{\partial x_1}, \qquad d = \frac{\partial X_2}{\partial x_2}, \tag{3.18}$$

all evaluated at (ξ, η). Thus in matrix form the linearization is $\dot{\boldsymbol{y}} = \boldsymbol{A}\boldsymbol{y}$, where

$$\boldsymbol{A} = \left. \begin{bmatrix} \dfrac{\partial X_1}{\partial x_1} & \dfrac{\partial X_1}{\partial x_2} \\ \dfrac{\partial X_2}{\partial x_1} & \dfrac{\partial X_2}{\partial x_2} \end{bmatrix} \right|_{(x_1, x_2) = (\xi, \eta)} \tag{3.19}$$

Example 3.2.3. Obtain the linearization of (3.9) by using the Taylor expansion of X_1 and X_2 about $(-1, 1)$.

Solution. Recall

$$X_1(x_1, x_2) = e^{x_1 + x_2} - x_2, \qquad X_2(x_1, x_2) = -x_1 + x_1 x_2.$$

The matrix $\boldsymbol{A}$ in (3.19) is

$$\boldsymbol{A} = \left. \begin{bmatrix} e^{x_1 + x_2} & e^{x_1 + x_2} - 1 \\ -1 + x_2 & x_1 \end{bmatrix} \right|_{(x_1, x_2) = (-1, 1)} = \begin{bmatrix} 1 & 0 \\ 0 & -1 \end{bmatrix}.$$

Therefore the linearization at $(-1, 1)$ is

$$\dot{y}_1 = y_1, \qquad \dot{y}_2 = -y_2;$$

in agreement with (3.14). □

3.3 The linearization theorem

This theorem relates the phase portrait of a non-linear system in the neighbourhood of a fixed point to that of its linearization.

Definition 3.3.1. A fixed point at the origin of a non-linear system $\dot{\mathbf{y}} = \mathbf{Y}(\mathbf{y})$, $\mathbf{y} \in S \subseteq \mathrm{R}^2$, is said to be *simple* if its linearized system is simple.

This definition extends the idea of simplicity (see Section 2.3) to the fixed points of non-linear systems. It can be used when the fixed point of interest is not at the origin by introducing local coordinates as in Section 3.2.

Theorem 3.3.1 (Linearization Theorem). Let the non-linear system

$$\dot{\mathbf{y}} = \mathbf{Y}(\mathbf{y}) \tag{3.20}$$

have a simple fixed point at $\mathbf{y} = \mathbf{0}$. Then, in a neighbourhood of the origin the phase portraits of the system and its linearization are qualitatively equivalent *provided* the linearized system is *not* a *centre.*

This important theorem is not proved in this book. The interested reader may consult Hartman (1964) if a proof is desired.

The linearization theorem is the basis of one of the main methods – 'linear stability analysis' – of investigating non-linear systems. However, it is easy to confuse the theorem and the techniques so that the significance of the theorem is not appreciated.

For example, one might, mistakenly, argue that the definition of the linearized system implies that, in a sufficiently small neighbourhood of the origin, the linear part of the vector field $\mathbf{Y}$ quantitatively approximates $\mathbf{Y}$ itself; and so their qualitative behaviour will obviously be the same. This is *not* a valid argument as the case when the linearization is a centre (excluded in the theorem) clearly shows.

Example 3.3.1. Show that the two systems

$$\dot{x}_1 = -x_2 + x_1(x_1^2 + x_2^2), \qquad \dot{x}_2 = x_1 + x_2(x_1^2 + x_2^2) \tag{3.21}$$

and

$$\dot{x}_1 = -x_2 - x_1(x_1^2 + x_2^2), \qquad \dot{x}_2 = x_1 - x_2(x_1^2 + x_2^2) \tag{3.22}$$

both have the same linearized systems at the origin, but that their phase portraits are qualitatively different.

Solution. Both systems (3.21) and (3.22) are already in the form (3.2), since both

$$\lim_{r \to 0} [x_1(x_1^2 + x_2^2)/r] \quad \text{and} \quad \lim_{r \to 0} [x_2(x_1^2 + x_2^2)/r]$$

are zero. Thus both systems have the linearization

$$\dot{x}_1 = -x_2, \qquad \dot{x}_2 = x_1 \tag{3.23}$$

which has a centre at $x_1 = x_2 = 0$. However, in polar coordinates (3.21) becomes

$$\dot{r} = r^3, \qquad \dot{\theta} = +1 \tag{3.24}$$

while (3.22) gives

$$\dot{r} = -r^3, \qquad \dot{\theta} = +1. \tag{3.25}$$

Equation (3.24) shows that $\dot{r} > 0$ for all $r > 0$ and so the trajectories of (3.21) spiral *outwards* as t increases. On the other hand, (3.25) has $\dot{r} < 0$ for all $r > 0$ and the trajectories of (3.22) spiral *inwards* (see Fig. 3.4). Thus (3.21) shows unstable behaviour while (3.22) is stable, i.e. they are qualitatively different. However, sufficiently close to the origin the vector fields of both (3.21) and (3.22) are quantitatively approximated, to whatever accuracy we choose, by the linear vector field in (3.23). □

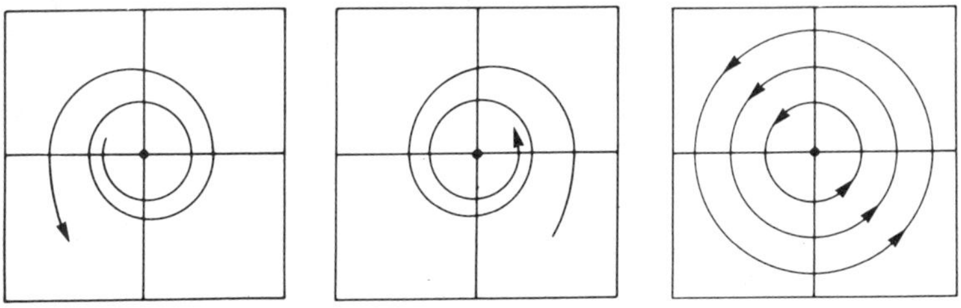

(a) (b) (c)

Fig. 3.4. Phase portraits for systems (a) (3.21); (b) (3.22) and (c) their linearization (3.23).

Example 3.3.1 shows that the quantitative proximity of **Y** and its linear part does not guarantee qualitative equivalence of the non-linear system and its linearization. The content of the linearization theorem is that the centre is the *only* exception. That is, provided the eigenvalues of the linearized system have non-zero real part, the phase portraits of the non-linear system and its linearization are qualitatively equivalent in a neighbourhood of the fixed point. Such fixed points are said to be *hyperbolic*. Some examples are shown in Fig. 3.5.

Observe that the qualitative equivalence is not confined to the ultimate classification of stable, saddle, unstable of Fig. 2.11. In fact, within the stable or unstable categories, nodes, improper nodes and foci in the linearization are preserved in the phase portrait of the non-linear system. We can therefore refer to the non-linear fixed points as nodes, foci, etc. when their linearizations are of that type.

This general property of hyperbolic fixed points arises from the special character of the continuous bijection involved in the qualitative equivalence. The mapping must reflect the quantitative agreement between **Y** and its linear part near the fixed point. On small enough neighbourhoods it must, in some sense, be close to the identity. This property of the continuous bijection allows us to deduce some more information from the linearized system.

A *separatrix* is a trajectory which approaches, or leaves, a fixed point tangent to a fixed radial direction. It follows that the tangents to the separatrices of the linearized system at the fixed point are also the tangents to the non-linear separatrices.

Example 3.3.2. Use the linearization theorem to determine the local phase portrait of the system

$$\dot{x}_1 = x_1 + 4x_2 + e^{x_1} - 1, \qquad \dot{x}_2 = -x_2 - x_2 e^{x_1} \tag{3.26}$$

at the origin.

Solution. The component functions of the vector field in (3.26) are twice differentiable and so (3.19) is applicable. We obtain

$$\boldsymbol{A} = \left.\begin{bmatrix} \dfrac{\partial}{\partial x_1}(x_1 + 4x_2 + e^{x_1} - 1) & \dfrac{\partial}{\partial x_2}(x_1 + 4x_2 + e^{x_1} - 1) \\ \dfrac{\partial}{\partial x_1}(-x_2 - x_2 e^{x_1}) & \dfrac{\partial}{\partial x_2}(-x_2 - x_2 e^{x_1}) \end{bmatrix}\right|_{(x_1, x_2) = (0,0)} \tag{3.27}$$

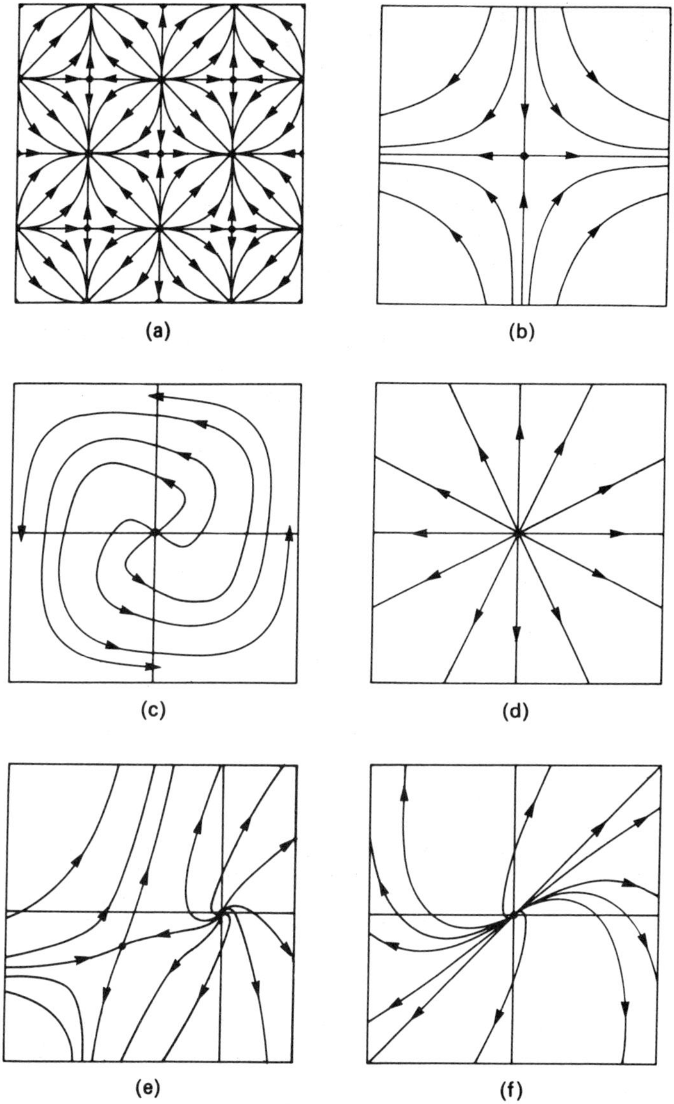

Fig. 3.5. Non-linear phase portraits and their linearizations at the origin: (a) non-linear system $\dot{x}_1 = \sin x_1$, $\dot{x}_2 = -\sin x_2$; (b) linearized system for (a) $\dot{x}_1 = x_1$, $\dot{x}_2 = -x_2$, a saddle point; (c) non-linear system $\dot{x}_1 = x_1 - x_2^3$, $\dot{x}_2 = x_2 + x_1^3$; (d) linearized system for (c) $\dot{x}_1 = x_1, \dot{x}_2 = x_2$, an unstable star node; (e) non-linear system $\dot{x}_1 = \frac{1}{2}(x_1 + x_2) + x_1^2, \dot{x}_2 = \frac{1}{2}(3x_2 - x_1)$; (f) linearized system for (e) $\dot{x}_1 = \frac{1}{2}(x_1 + x_2)$, $\dot{x}_2 = \frac{1}{2}(3x_2 - x_1)$. All these fixed points are hyperbolic.

Hence the linearization is

$$\dot{x}_1 = 2x_1 + 4x_2, \qquad \dot{x}_2 = -2x_2, \tag{3.28}$$

which is a simple linear system with a saddle point at the origin ($\det \boldsymbol{A} < 0$).

The directions of the separatrices are given by the eigenvectors $(1, 0)$ and $(1, -1)$ of $\boldsymbol{A}$. The corresponding eigenvalues are $+2$ (unstable) and -2 (stable), respectively. Therefore, the unstable separatrices lie on the line $x_2 = 0$ while the stable ones lie on $x_2 = -x_1$ (see Fig. 3.6(b)).

For (3.26), there is a neighbourhood of the origin in which the non-linear separatrices are as shown in Fig. 3.6(a). This follows because, on the line $x_2 = -x_1$, $dx_2/dx_1 \gtrless -1$ when $x_1 \lessgtr 0$. □

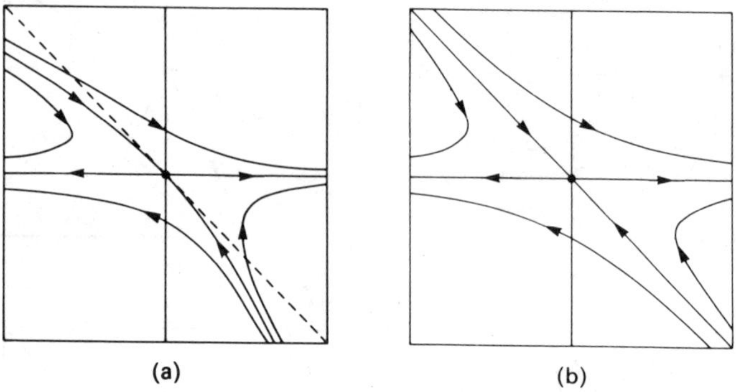

Fig. 3.6. Phase portraits for (a) the non-linear system (3.26) and (b) its linearized system (3.28). Notice that the x_1-axis is the unstable separatrix for both systems because both (3.26) and (3.28) imply $\dot{x}_2 = 0$ when $x_2 = 0$. The stable separatrices of (3.26) and (3.28), however, only become tangential at the fixed point.

Further examples of the tangential relation between separatrices in non-linear systems and their linearizations appear in Fig. 3.5. The star node in items (c) and (d) of this figure is particularly interesting because every trajectory of both linear and non-linear systems is a separatrix. Thus every non-linear trajectory ultimately becomes tangential to its radial counterpart in the linearized system.

Our examples show that in analysing non-linear systems it is the directions of the straight line separatrices in the linearization that are important. They provide us with the direction of the non-linear separatrices at the fixed point. These directions are referred to as the *principal directions* at the fixed point and they are usually obtained (as in Example 3.3.2) from the linear transformation relating the linearized system to its canonical system.

3.4 Non-simple fixed points

A fixed point of a non-linear system is said to be *non-simple* if the corresponding linearized system is non-simple. Recall that such linear systems contain a straight line, or possibly a whole plane, of fixed points. The non-linear terms in g_1 and g_2 can drastically modify this behaviour; see Fig. 3.7 for example.

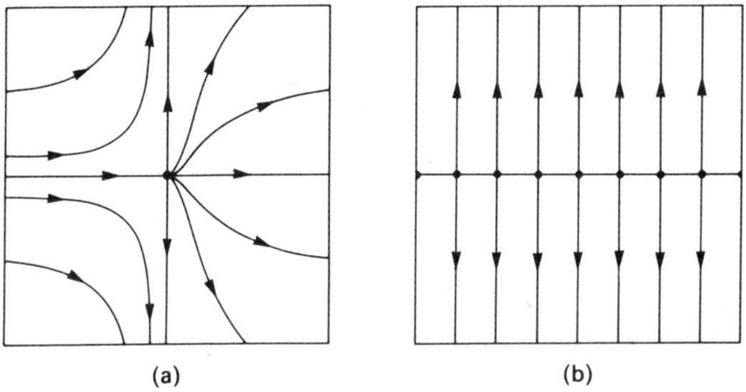

Fig. 3.7. Phase portraits for (a) the system $\dot{x}_1 = x_1^2, \dot{x}_2 = x_2$ and (b) its linearization at $(0, 0)$: $\dot{x}_1 = 0$, $\dot{x}_2 = x_2$, where the x_1-axis is a line of fixed points.

The nature of the local phase portrait is now determined by non-linear terms. Therefore, in contrast to the simple fixed points of Section 3.3, there are infinitely many different types of local phase portrait. Some examples of what can occur for just low order polynomials in x_1 and x_2 are shown in Figs. 3.8–3.10. The linearizations of all of the systems shown in these figures have at least a line of fixed points as their phase portraits. These diagrams illustrate comparatively straightforward non-linear vector fields.

Lines of fixed points can also occur in non-linear systems; they are not necessarily straight and *always* consist of non-simple fixed points.

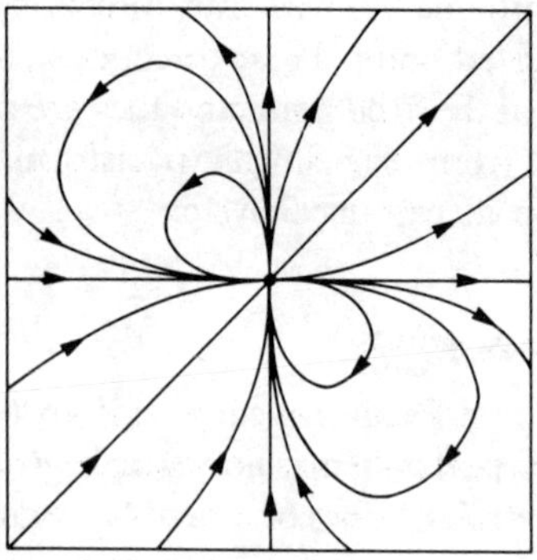

Fig. 3.8. $\dot{x}_1 = x_1(x_1+2x_2)$, $\dot{x}_2 = x_2(2x_1+x_2)$; linearization $\dot{x}_1 = 0$, $\dot{x}_2 = 0$.

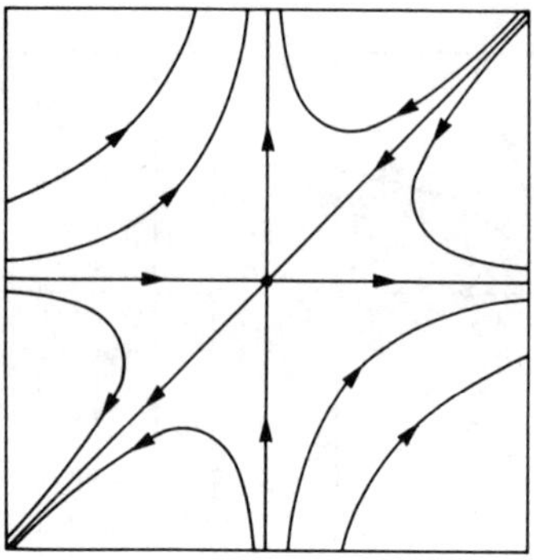

Fig. 3.9. $\dot{x}_1 = x_1(x_1-2x_2)$, $\dot{x}_2 = -x_2(2x_1-x_2)$; linearization $\dot{x}_1 = 0$, $\dot{x}_2 = 0$.

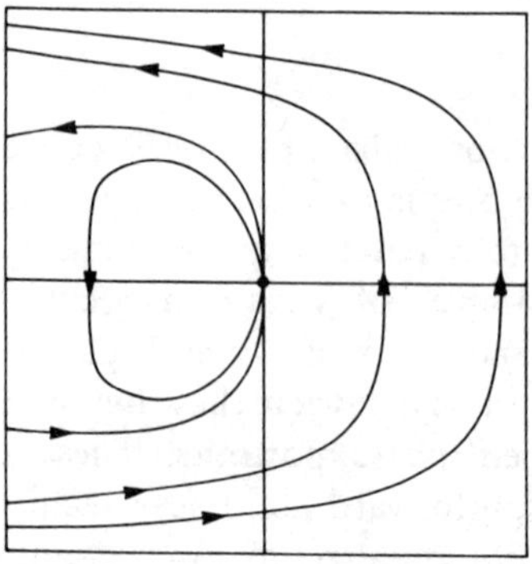

Fig. 3.10. $\dot{x}_1 = -x_2^5$, $\dot{x}_2 = x_1+x_2^2$; linearization $\dot{x}_1 = 0$, $\dot{x}_2 = x_1$.

Consider the system

$$\dot{x}_1 = x_1 - x_2^2, \qquad \dot{x}_2 = x_2(x_1 - x_2^2), \tag{3.29}$$

for example. The fixed points of (3.29) lie on the parabola $x_2^2 = x_1$. A typical fixed point (k^2, k), k real, has linearization $\dot{\boldsymbol{y}} = \boldsymbol{A}\boldsymbol{y}$, where

$$\boldsymbol{A} = \begin{bmatrix} 1 & -2k \\ k & -2k^2 \end{bmatrix}. \tag{3.30}$$

Clearly, det $(\boldsymbol{A}) = 0$ for all k and so the fixed points are all non-simple. The phase portrait for (3.29) is shown in Fig. 3.11.

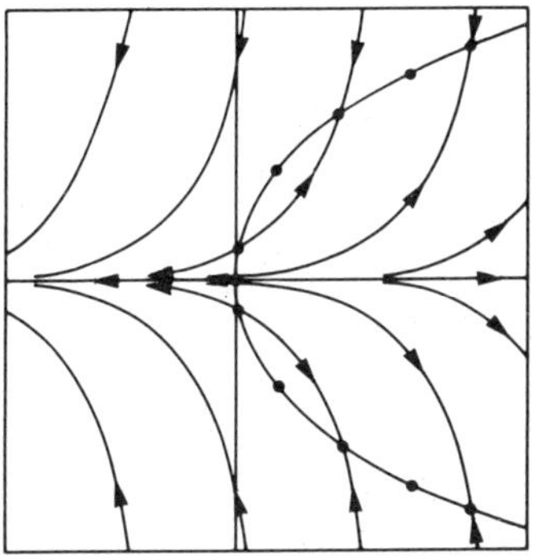

Fig. 3.11. The phase portrait for the system $\dot{x}_1 = x_1 - x_2^2, \dot{x}_2 = x_2(x_1 - x_2^2)$ with non-simple fixed points on the parabola $x_2^2 = x_1$.

In view of the above observations, it is not surprising that there is no detailed classification of non-simple fixed points. However, the following definitions of stability (which apply to both simple and non-simple fixed points) do provide a coarse classification of qualitative behaviour.

3.5 Stability of fixed points

It can be shown that the local phase portrait in the neighbourhood of any fixed point falls into one, and only one, of three stability types: asymptotically stable, neutrally stable or unstable.

Definition 3.5.1. A fixed point $\mathbf{x}_0$ of the system $\dot{\mathbf{x}} = \mathbf{X}(\mathbf{x})$ is said to be *asymptotically stable* if there is a neighbourhood N of $\mathbf{x}_0$ such that every trajectory passing through N approaches $\mathbf{x}_0$ as t tends to infinity.

This is the type of stability encountered in connection with the node, improper node and focus in Section 2.3. Obviously, Definition 3.5.1 can be applied to simple, non-linear fixed points via the linearized system. For example, the non-linear system

$$\dot{x}_1 = -x_1 + x_2 - x_1^3, \qquad \dot{x}_2 = -x_1 - x_2 + x_2^2 \tag{3.31}$$

has an asymptotically stable fixed point at the origin. This follows because the linearized system

$$\dot{x}_1 = -x_1 + x_2, \qquad \dot{x}_2 = -x_1 - x_2 \tag{3.32}$$

has eigenvalues $-1 \pm i$, so that the origin is a stable focus. The existence of a neighbourhood of the kind required by Definition 3.5.1, in the phase plane of the system (3.31), is then assured by the linearization theorem.

Definition 3.5.1 can also be applied to non-simple, non-linear fixed points. For example, the non-linear system

$$\dot{x}_1 = -x_1 x_2^2, \qquad \dot{x}_2 = -x_2 x_1^2 \tag{3.33}$$

has a non-simple fixed point at the origin. In polar coordinates, however,

$$\dot{r} = -2x_1^2 x_2^2 / r < 0$$

for any (x_1, x_2) in the phase plane. Thus, all trajectories approach the origin as $t \to \infty$, so that Definition 3.5.1 is satisfied with $N = \mathbb{R}^2$.

The second type of stability is weaker than asymptotic stability.

Definition 3.5.2. A fixed point $\mathbf{x}_0$ of the system $\dot{\mathbf{x}} = \mathbf{X}(\mathbf{x})$ is said to be *stable* if, for every neighbourhood N of $\mathbf{x}_0$, there is a smaller neighbourhood $N' \subseteq N$ of $\mathbf{x}_0$ such that every trajectory which passes through N' remains in N as t increases.

Observe that every asymptotically stable fixed point is stable by taking $N' = N$. However, the converse is not true.

Example 3.5.1. Show that the system

$$\dot{x}_1 = x_2, \qquad \dot{x}_2 = -x_1^3 \tag{3.34}$$

is stable at the origin but not asymptotically stable.

Solution. The fixed point at the origin of (3.34) is non-simple (the linearized system is $\dot{x}_1 = x_2, \dot{x}_2 = 0$) so that the linearization theorem does not provide a local phase portrait. However, the shape of the trajectories is given by

$$\frac{\mathrm{d}x_2}{\mathrm{d}x_1} = -\frac{x_1^3}{x_2} \tag{3.35}$$

which has solutions satisfying

$$\tfrac{1}{2}x_1^4 + x_2^2 = C, \tag{3.36}$$

where C is a real constant. The phase portrait is shown in Fig. 3.12.

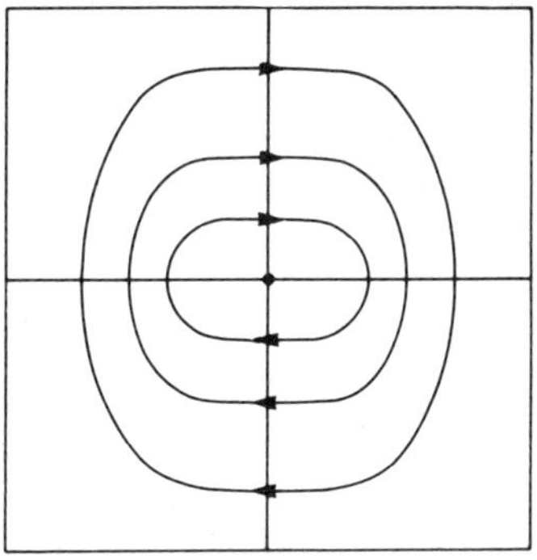

Fig. 3.12. Phase portrait for the system (3.34). Trajectories satisfy $\frac{1}{2}x_1^4 + x_2^2 = C$. Orientation given by $\dot{x}_1 > 0$ for $x_2 > 0$.

None of the trajectories approach the origin as $t \to \infty$, so the fixed point is not asymptotically stable. However, as Fig. 3.13 shows, for every disc N centred on the origin, there is a smaller disc N' such that every trajectory passing through N' remains in N. Thus the origin is stable. □

Definition 3.5.3. A fixed point of the system $\dot{\mathbf{x}} = \mathbf{X}(\mathbf{x})$ which is stable but not asymptotically stable is said to be *neutrally stable.*

There are many examples of neutrally stable fixed points similar to Example 3.5.1. For instance, the non-trivial fixed point of the Volterra–Lotka equations

$$\dot{x}_1 = x_1(a - bx_2), \qquad \dot{x}_2 = -x_2(c - dx_1), \tag{3.37}$$

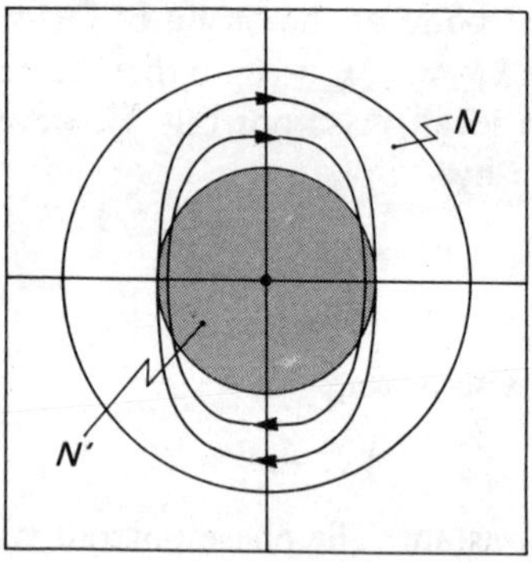

Fig. 3.13. Typical neighbourhoods N and N' (shaded) of Definition 3.5.2. Observe all trajectories passing through N' remain in N.

$a, b, c, d > 0$ is neutrally stable. The phase portrait for these equations was given in Fig. 1.33. Neutral stability of the fixed point at $(c/d, a/b)$ follows from the existence of neighbourhoods N and N' satisfying Definition 3.5.2 as indicated in Fig. 3.14. Clearly, the fixed point is not asymptotically stable.

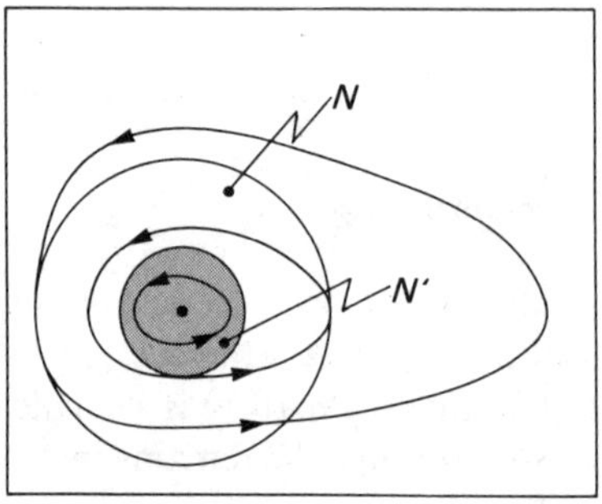

Fig. 3.14. Typical neighbourhoods N and N' for the Volterra–Lotka equations showing neutral stability.

Another example is the non-simple linear fixed point shown in Fig. 3.15. The particular fixed point A is not asymptotically stable because there are trajectories passing through N (see Fig. 3.15) which do not approach A as $t \to \infty$. However, with $N' = N$, every trajectory passing through N' remains in N; so A is stable.

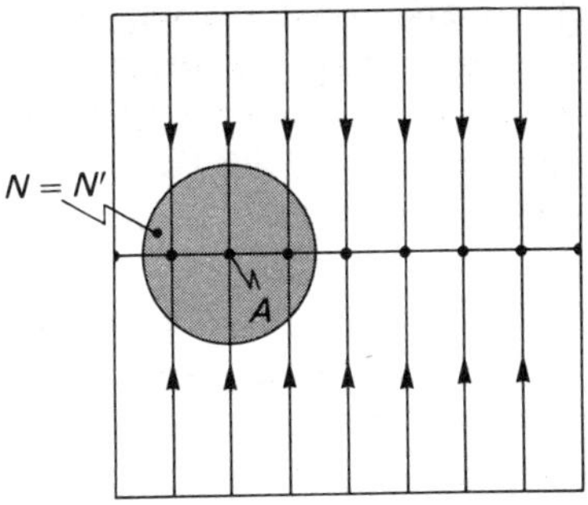

Fig. 3.15. Neutral stability of the system $\dot{x}_1 = 0$, $\dot{x}_2 = -x_2$ at A follows with $N = N'$.

Definition 3.5.4. A fixed point of the system $\dot{\mathbf{x}} = \mathbf{X}(\mathbf{x})$ which is not stable is said to be *unstable.*

This means that there is a neighbourhood N of the fixed point such that for every neighbourhood $N' \subseteq N$ there is at least one trajectory which passes through N' and does *not* remain in N. For example, the saddle point is unstable because there is a separatrix, containing points arbitrarily close to the origin, which escapes to infinity with increasing t.

Example 3.5.2. What are the stability types of the simple linear fixed points obtained in Section 2.3.

Solution. (a) Stable node, improper node and focus are such that all trajectories approach the origin as $t \to \infty$. Thus Definition 3.5.1 is satisfied with $N = \mathrm{R}^2$ and these fixed points are asymptotically stable.

(b) The centre is not asymptotically stable but it is stable. This follows from Definition 3.5.2 with $N' = N = \{(x_1, x_2) | (x_1^2 + x_2^2)^{1/2} < \varepsilon, \varepsilon > 0\}$. Thus the centre is neutrally stable.

(c) The unstable node, improper node, focus and saddle point are all unstable in the sense of Definition 3.5.4. □

3.6 Ordinary points and global behaviour

3.6.1 *Ordinary points*

Any point in the phase plane of the system $\dot{\mathbf{x}} = \mathbf{X}(\mathbf{x})$ which is not a fixed point is said to be an *ordinary point*. Thus, if $\mathbf{x}_0$ is an ordinary point then $\mathbf{X}(\mathbf{x}_0) \neq \mathbf{0}$ and, by the continuity of $\mathbf{X}$, there is a neighbourhood of $\mathbf{x}_0$ containing only ordinary points. This means that the local phase portrait at an ordinary point has no fixed points. There is an important result concerning the qualitative equivalence of such local phase portraits – the *flow box theorem* (Hirsch and Smale, 1974).

Consider the local phase portraits at a typical ordinary point $\mathbf{x}_0$ shown in Figs. 3.16–3.19. In each case, a special neighbourhood of $\mathbf{x}_0$, called a flow box, is shown. The trajectories of the system enter at one end and flow out through the other; no trajectories leave through the sides. For each phase portrait shown, we can find new coordinates for the plane such that the local phase portrait in the flow box looks like the one shown in Fig. 3.16. For example, in Fig. 3.17 we take polar coordinates (r, θ). In the r, θ-plane the circles (r = constant) become straight lines parallel to the $\theta = 0$ axis and the radial lines (θ = constant) become straight lines parallel to the $r = 0$ axis. Thus, the phase portrait in the flow box in Fig. 3.17 is, in the r, θ-plane, the same as Fig. 3.16.

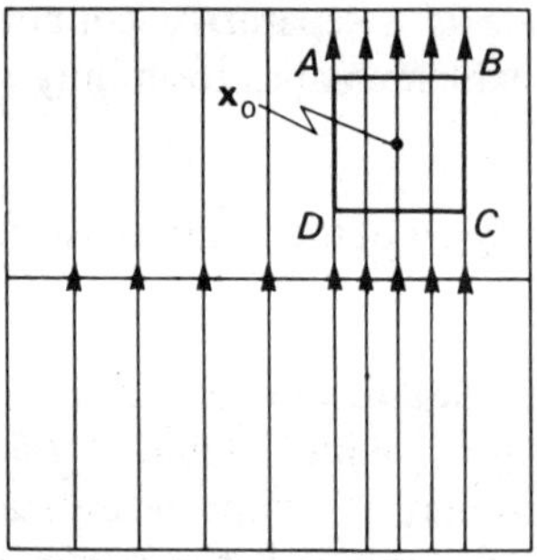

Fig. 3.16. System $\dot{x}_1 = 0$, $\dot{x}_2 = 1$ with typical flow box.

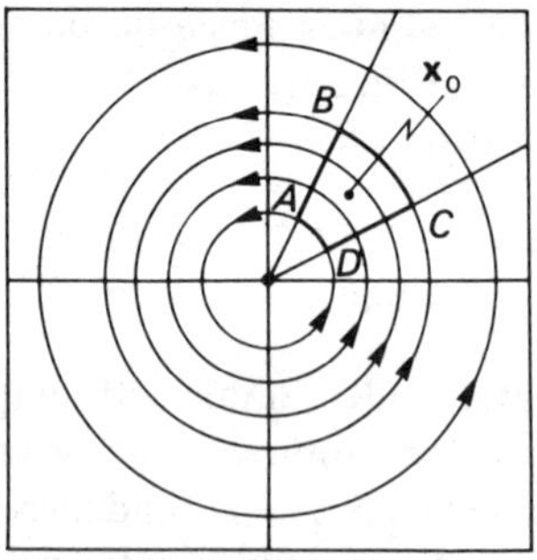

Fig. 3.17. System $\dot{x}_1 = -x_2$, $\dot{x}_2 = x_1$. In polar coordinates this gives $\dot{r} = 0$, $\dot{\theta} = 1$.

For Fig. 3.18, the trajectories in the neighbourhood of $\mathbf{x}_0$ lie on hyperbolae $x_1 x_2 = K > 0$. If we introduce variables $y_1 = x_1 x_2$ and $y_2 = \ln x_1$ then the flow box is bounded by the coordinate lines y_1 = constant and y_2 = constant and in the $y_1 y_2$-plane the local phase portrait again looks like Fig. 3.16.

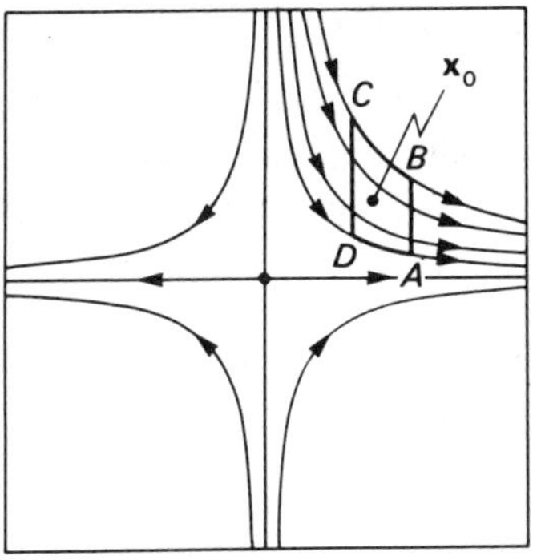

Fig. 3.18. System $\dot{x}_1 = x_1, \dot{x}_2 = -x_2$. The variables $y_1 = x_1 x_2$ and $y_2 = \ln x_1$, $x_1 > 0$, satisfy $\dot{y}_1 = 0, \dot{y}_2 = 1$.

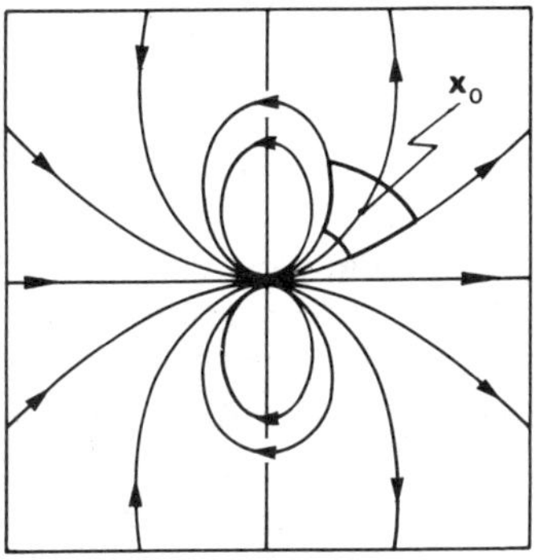

Fig. 3.19. The flow box theorem guarantees the existence of coordinates which transform the local phase portrait at $\mathbf{x}_0$ into the form shown in Fig. 3.16.

Theorem 3.6.1 (Flow Box Theorem). In a sufficiently small neighbourhood of an ordinary point $\mathbf{x}_0$ of the system $\dot{\mathbf{x}} = \mathbf{X}(\mathbf{x})$ there is a differentiable change of coordinates $\mathbf{y} = \mathbf{y}(\mathbf{x})$ such that $\dot{\mathbf{y}} = (0, 1)$.

The flow box theorem guarantees the *existence* of new coordinates with the above property, at least in some neighbourhood of *any* ordinary point of *any* system. Thus, local phase portraits at ordinary points are all qualitatively equivalent.

3.6.2 *Global phase portraits*

The linearization and flow box theorems provide *local* phase portraits at most simple fixed points and all ordinary points. However, this information does not always determine the complete phase portrait of a system.

Example 3.6.1. Find and classify the fixed points of the system

$$\dot{x}_1 = 2x_1 - x_1^2, \qquad \dot{x}_2 = -x_2 + x_1 x_2. \tag{3.38}$$

Discuss possible phase portraits for the system.

Solution. The system has fixed points at $A = (0,0)$ and $B = (2,0)$. The linearized systems are:

$$\dot{x}_1 = 2x_1, \qquad \dot{x}_2 = -x_2 \quad \text{at } A; \tag{3.39}$$

and

$$\dot{y}_1 = -2y_1, \qquad \dot{y}_2 = y_2 \quad \text{at } B. \tag{3.40}$$

The linearization theorem implies that (3.38) has saddle points at A and B. Furthermore, the non-linear separatrices of these saddle points are tangent to the principal directions at A and B. For (3.39) and (3.40) the principal directions coincide with the local coordinate axes.

This information is insufficient to determine the qualitative type of the global phase portrait. For example, Fig. 3.20 shows two phase

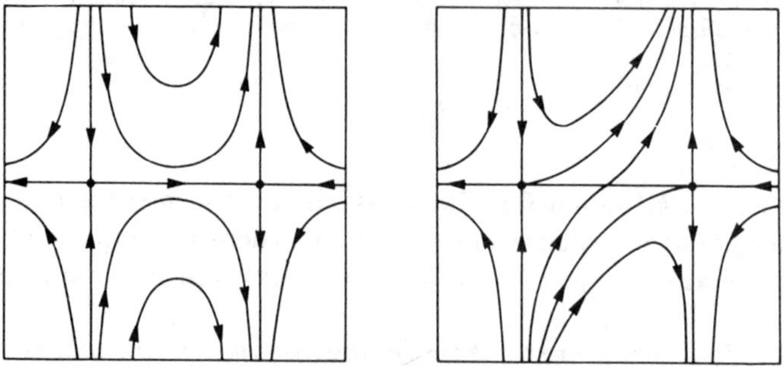

(a) (b)

Fig. 3.20. Two qualitatively different phase portraits compatible with the local phase portraits obtained from the linearization theorem.

portraits that are consistent with the local behaviour. The phase portraits shown are not qualitatively equivalent because the two saddle points have a *common separatrix* or 'saddle connection' in Fig. 3.20(a) whereas in Fig. 3.20(b) they do not. This is a qualitative difference; there is no continuous bijection that relates the two phase portraits.

Returning to (3.38), observe that $\dot{x}_1 \equiv 0$ on the lines $x_1 = 0$ and $x_1 = 2$ so that trajectories coincide with these lines. Furthermore, $\dot{x}_2 = 0$ when $x_2 = 0$. These observations indicate that Fig. 3.20(a) gives the correct qualitative type of phase portrait. □

3.7 First integrals

Definition 3.7.1. A continuously differentiable function f: $D(\subseteq \mathsf{R}^2) \to \mathsf{R}$ is said to be a *first integral* of the system $\dot{\mathbf{x}} = \mathbf{X}(\mathbf{x})$, $\mathbf{x} \in S \subseteq \mathsf{R}^2$ on the region $D \subseteq S$ if $f(\mathbf{x}(t))$ is constant for any solution $\mathbf{x}(t)$ of the system.

When a first integral exists it is not unique. Clearly if $f(\mathbf{x})$ is a first integral then so is $f(\mathbf{x}) + C$ or $Cf(\mathbf{x})$, C real. The constant C in the former is often chosen to provide a convenient value of the first integral at $\mathbf{x} = \mathbf{0}$. The trivial first integral which is *identically* constant on D, is always excluded from our considerations.

The fact that f is a first integral for $\dot{\mathbf{x}} = \mathbf{X}(\mathbf{x})$ can be expressed in terms of its (continuous) partial derivatives $f_{x_1} \equiv \partial f/\partial x_1, f_{x_2} \equiv \partial f/\partial x_2$. Since f is constant on any solution $\mathbf{x}(t) = (x_1(t), x_2(t))$,

$$\frac{\mathrm{d}}{\mathrm{d}t} f(\mathbf{x}(t)) = 0 = \dot{x}_1(t) f_{x_1}(\mathbf{x}(t)) + \dot{x}_2 f_{x_2}(\mathbf{x}(t)) \tag{3.41}$$

$$= X_1(\mathbf{x}(t)) f_{x_1}(\mathbf{x}(t)) + X_2(\mathbf{x}(t)) f_{x_2}(\mathbf{x}(t)) \tag{3.42}$$

$$= \lim_{h \to 0} \left\{ \frac{f(\mathbf{x} + h\mathbf{X}(\mathbf{x})) - f(\mathbf{x})}{h} \right\} \Bigg|_{\mathbf{x} = \mathbf{x}(t)}. \tag{3.43}$$

Equation (3.43) holds at every point of D, so that the directional derivative of f along the vector field $\mathbf{X}$, is identically zero on D.

A first integral is useful because of the connection between its *level curves* (defined by $f(\mathbf{x}) =$ constant) and the trajectories of the system. Consider the level curve $L_C = \{\mathbf{x} | f(\mathbf{x}) = C\}$. Let $\mathbf{x}_0 \in L_C$ and let $\xi(t)$ be the trajectory passing through the point $\mathbf{x}_0$ in the phase plane.

Since f is first integral, $f(\xi(t))$ is constant and $f(\xi(t)) = f(\mathbf{x}_0) = C$. Therefore, the trajectory passing through $\mathbf{x}_0$ lies in L_C.

When f is a first integral, it is constant on every trajectory in D. Thus every trajectory is part of some level curve of f. Hence each level curve is a union of trajectories. Uniqueness of the solutions to $\dot{\mathbf{x}} = \mathbf{X}(\mathbf{x})$ ensures that this union is disjoint.

The first integral is so named because it is usually obtained from a single integration of the differential equation

$$\frac{\mathrm{d}x_2}{\mathrm{d}x_1} = \frac{X_2(x_1, x_2)}{X_1(x_1, x_2)}, \qquad (x_1, x_2) \in D' \subseteq S. \tag{3.44}$$

If the solutions of this equation satisfy

$$f(x_1, x_2) = C, \tag{3.45}$$

where $f\colon D' \to \mathsf{R}$, then f is first integral of $\dot{\mathbf{x}} = \mathbf{X}(\mathbf{x})$ on D'. This follows by differentiating (3.45),

$$\frac{\mathrm{d}f}{\mathrm{d}x_1} \equiv 0 = f_{x_1} + \frac{\mathrm{d}x_2}{\mathrm{d}x_1} f_{x_2},$$

substituting from (3.44) and multiplying by $X_1(x_1, x_2)$ to obtain (3.42). Of course, X_1 cannot vanish on D' or (3.44) would not define $\mathrm{d}x_2/\mathrm{d}x_1$. However, zeroes of X_1 present no such problem in (3.42). Thus, if $f(\mathbf{x})$ is continuously differentiable on a larger set $D \supset D'$ and (3.42) is satisfied there, then f is a first integral of $\dot{\mathbf{x}} = \mathbf{X}(\mathbf{x})$ on D.

Definition 3.7.2. A system that has a first integral on the whole of the plane (i.e. $D = \mathsf{R}^2$) is said to be *conservative*.

Example 3.7.1. Show that the system

$$\dot{x}_1 = -x_2, \qquad \dot{x}_2 = x_1 \tag{3.46}$$

is conservative, while the system

$$\dot{x}_1 = x_1, \qquad \dot{x}_2 = x_2 \tag{3.47}$$

is not.

Solution. The differential equation (3.44) gives

$$\frac{\mathrm{d}x_2}{\mathrm{d}x_1} = -\frac{x_1}{x_2}, \qquad x_2 \neq 0 \tag{3.48}$$

which has solutions satisfying

$$x_1^2+x_2^2=C, \qquad x_2 \neq 0, \tag{3.49}$$

where C is a positive constant. However, (3.42) with

$$f(\mathbf{x}) = x_1^2+x_2^2 \tag{3.50}$$

is satisfied for *all* $(x_1, x_2) \in \mathrm{R}^2$. Thus (3.50) is a first integral of the system (3.46) on the whole plane and so (3.46) is a conservative system.

Now consider (3.47); the differential equation

$$\frac{dx_2}{dx_1} = \frac{x_2}{x_1}, \qquad x_1 \neq 0 \tag{3.51}$$

has solutions

$$x_2 = Cx_1 \tag{3.52}$$

with C real. In this case (3.42) is satisfied by

$$f(\mathbf{x}) = x_2/x_1, \qquad x_1 \neq 0 \tag{3.53}$$

so that D' is R^2 less the x_2-axis.

There is no way in which the domain D' of f can be enlarged upon. The only continuous function which is:

(a) defined on the whole x_1x_2-plane;
(b) constant on every trajectory of (3.47) (i.e. on every radial line in the plane and at the origin itself);

is identically constant. This follows because on any radial line we can easily find a sequence of points $\{\mathbf{x}_i\}_{i=1}^{\infty}$ such that $\lim_{i\to\infty} \mathbf{x}_i = \mathbf{0}$. Thus, by continuity, $f(\mathbf{x}_i) = f(\mathbf{0})$ for all i and f takes the same value at all points of all radial lines. In other words, (a) and (b) are only satisfied by a function which is constant throughout R^2. Thus, there is no first integral on R^2 and (3.47) is not a conservative system. □

Conservative systems play an important part in mechanical problems. The equations of motion of such systems can be constructed from their *Hamiltonian*. For example, a particle moving in one dimension, with position coordinate x, momentum p and Hamiltonian $H(x, p)$ has equations of motion

$$\dot{x} = \frac{\partial H(x,p)}{\partial p}, \qquad \dot{p} = -\frac{\partial H(x,p)}{\partial x}. \tag{3.54}$$

Here $H(x, p)$ is a first integral for (3.54) because

$$\dot{x}\frac{\partial H}{\partial x} + \dot{p}\frac{\partial H}{\partial p} = \frac{\mathrm{d}}{\mathrm{d}t}H \equiv 0$$

(cf. (3.42)) and H remains constant along the trajectories. In other words, H is a conserved quantity or constant of the motion.

Example 3.7.2. Find the Hamiltonian H of the system

$$\dot{x} = p, \qquad \dot{p} = -x + x^3 \tag{3.55}$$

and sketch its phase portrait.

Solution. The differential equation

$$\frac{\mathrm{d}p}{\mathrm{d}x} = \frac{-x + x^3}{p}, \qquad p \neq 0 \tag{3.56}$$

has solutions on the plane less the x-axis which satisfy

$$x^2 - \tfrac{1}{2}x^4 + p^2 = C, \tag{3.57}$$

where C is a constant. Equation (3.42) shows that this is a first integral on the whole of the plane and $H(x, p) = x^2 - \frac{1}{2}x^4 + p^2$.

The level curves of the first integral are unions of trajectories of (3.55), so we can obtain the global phase portrait for the system by sketching the level curves of $H(x, p)$. These curves are shown in Fig. 3.21(a) and the phase portrait in Fig. 3.21(b). □

We can show that a given level curve is a union of several trajectories by considering $L_{1/2} = \{\mathbf{x} \mid x^2 - \frac{1}{2}x^4 + p^2 = \frac{1}{2}\}$ for $H(x, p)$. This set of points is shown as a heavy line in Fig. 3.21(a). In Fig. 3.21(b), it is made up from eight trajectories.

Notice also that the linearization of (3.55) at the origin is a centre, so that the linearization theorem could not provide a local phase portrait there. In fact, first integrals are one of the main ways of detecting centres in non-linear systems.

Example 3.7.3. Show that the fixed point of the system

$$\dot{x}_1 = x_1 - x_1x_2, \qquad \dot{x}_2 = -x_2 + x_1x_2 \tag{3.58}$$

at (1, 1) is a centre.

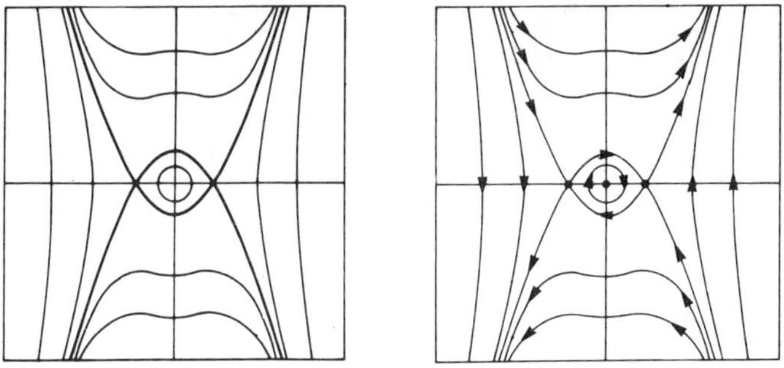

Fig. 3.21. (a) Level curves of $H(x,p) = x^2 - \frac{1}{2}x^4 + p^2$. $L_{1/2}$ is marked by a heavy line. (b) Trajectories of (3.55). The orientations are determined by noting that $\dot{x} > 0$ for $p > 0$ and $\dot{x} < 0$ for $p < 0$.

Solution. The differential equation

$$\frac{dx_2}{dx_1} = \frac{-x_2 + x_1 x_2}{x_1 - x_1 x_2}, \qquad x_1 \neq x_1 x_2, \tag{3.59}$$

is separable, with solutions satisfying

$$g(x_1)g(x_2) = K, \tag{3.60}$$

where $g(x) = xe^{-x}$ and K is a positive constant. The function g is shown in Fig. 3.22 for $x \geqslant 0$; it has a single maximum at $x = 1$ where $g(1) = e^{-1}$. It follows that $(x_1, x_2) = (1, 1)$ is a maximum of the first integral $g(x_1)g(x_2)$. This means that there is a neighbourhood of (1, 1) in which the level curves of $g(x_1)g(x_2)$ are *closed*. Since, these level curves coincide with trajectories, we conclude that (1, 1) is a centre. □

It is important to realize that first integrals do not give solutions $\mathbf{x}(t)$ for a system; rather they provide the shape of the trajectories.

Example 3.7.4. Show that the systems

$$\dot{x}_1 = x_1, \qquad \dot{x}_2 = -x_2 \tag{3.61}$$

and

$$\dot{x}_1 = x_1(1 - x_2), \qquad \dot{x}_2 = -x_2(1 - x_2) \tag{3.62}$$

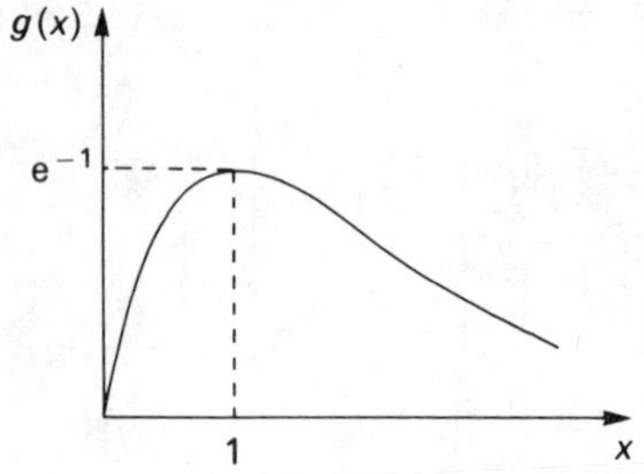

Fig. 3.22. Plot xe^{-x} versus x for $x \geqslant 0$.

have the same first integral and sketch their phase portraits.

Solution. The trajectories of both systems lie on the solutions of

$$\frac{dx_2}{dx_1} = -\frac{x_2}{x_1}, \qquad x_1 \neq 0, \tag{3.63}$$

and both have

$$f(\mathbf{x}) = x_1 x_2 \tag{3.64}$$

as a first integral on R^2. The level curves of f are rectangular hyperbolae which, for (3.61), can be oriented by noting the direction of $\dot{\mathbf{x}}$ on the coordinate axes. This is simply the linear saddle familiar from Section 2.3.

The system (3.62) has fixed points at the origin and everywhere on the line $x_2 = 1$. Furthermore, $\dot{x}_2 > 0$ for $x_2 > 1$ and for $x_2 < 0$ while $\dot{x}_2 < 0$ for $0 < x_2 < 1$. Therefore the phase portrait must be that shown in Fig. 3.23. □

3.8 Limit cycles

Definition 3.8.1. A closed trajectory C in a phase portrait is called a *limit cycle* if it is isolated from all other closed trajectories; more precisely, if there is a tubular neighbourhood of C which contains no other closed trajectories.

We can illustrate this definition by contrasting the limit cycle with the centre (see Fig. 3.24). It is easy to construct examples of limit cycles in polar coordinates.

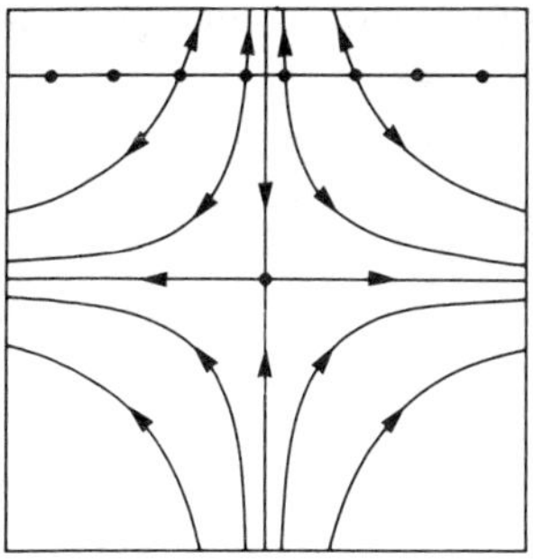

Fig. 3.23. Phase portrait for system (3.62). This system has the same first integral as the linear saddle.

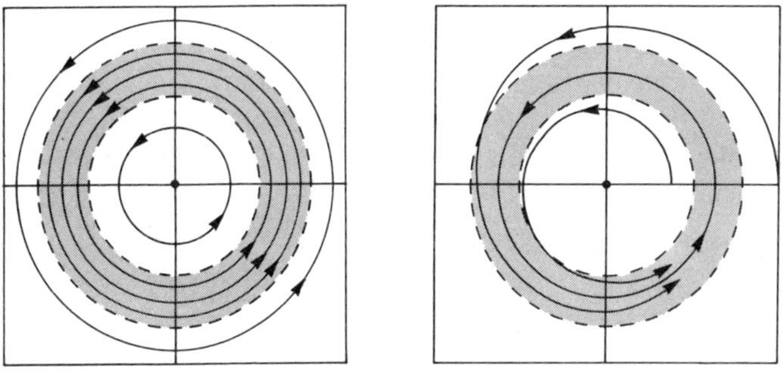

(a) (b)

Fig. 3.24. Illustration of a tubular neighbourhood, here taken as a circular annulus (shaded). Observe that for a centre, shown in (a), closed orbits are not isolated. In (b), however, the limit cycle is the only closed orbit in the tubular neighbourhood.

Example 3.8.1. Show that the system

$$\dot{x}_1 = -x_2 + x_1[1 - (x_1^2 + x_2^2)^{1/2}],$$
$$\dot{x}_2 = x_1 + x_2[1 - (x_1^2 + x_2^2)^{1/2}] \tag{3.65}$$

has a limit cycle given by $x_1^2 + x_2^2 = 1$.

Solution. In polar coordinates $x_1 = r\cos\theta$, $x_2 = r\sin\theta$, (3.65) becomes

$$\dot{r} = r(1-r), \qquad \dot{\theta} = 1. \tag{3.66}$$

Clearly, $r(t) \equiv 1$ and $\theta(t) = t$ is a solution giving a closed trajectory consisting of the circle $x_1^2 + x_2^2 = 1$ traversed in an anticlockwise sense at constant angular speed $\dot{\theta} = 1$. For $0 < r < 1$, $\dot{r}$ is positive and trajectories in this region spiral outwards (recall $\dot{\theta} = 1$) towards $r = 1$. However, $r > 1$, gives $\dot{r} < 0$ and trajectories must spiral inwards as t increases. The phase portrait is qualitatively equivalent to that shown in Fig. 3.24(b) with a limit cycle given by $x_1^2 + x_2^2 = 1$. □

Limit cycles do not all behave in the same way as the one in Example 3.8.1. There are three types:

(a) the *stable* (or *attracting*) limit cycle, where trajectories spiral into the closed orbit from both sides as $t \to \infty$, i.e. as in Fig. 3.24(b);
(b) the *unstable* (or *repelling*) limit cycle, where trajectories spiral away from the closed orbit on both sides as $t \to \infty$;
(c) the *semi-stable* limit cycle where the trajectories spiral towards the closed trajectory from one side and spiral away from it on the other.

Example 3.8.2. Find the limit cycles in the following systems and give their types:

(a) $\dot{r} = r(r-1)(r-2), \quad \dot{\theta} = 1;$ (3.67)
(b) $\dot{r} = r(r-1)^2, \quad \dot{\theta} = 1.$ (3.68)

Solution
(a) There are closed trajectories given by

$$r(t) \equiv 1, \quad \theta = t \quad \text{and} \quad r(t) \equiv 2, \quad \theta = t. \tag{3.69}$$

Furthermore

$$\dot{r} \begin{cases} > 0, & 0 < r < 1 \\ < 0, & 1 < r < 2 \\ > 0, & r > 2 \end{cases}. \tag{3.70}$$

The system therefore has two circular limit cycles: one stable ($r = 1$) and one unstable ($r = 2$).

(b) System (3.68) has a single circular limit cycle of radius one. However, $\dot{r}$ is positive for $0 < r < 1$ *and* $r > 1$, so the limit cycle is semistable. □

Limit cycles are not always circular and are, therefore, not always revealed by simply changing to polar coordinates. For example, consider the 'Van der Pol' equation

$$\ddot{x} - \dot{x}(1 - x^2) + x = 0 \tag{3.71}$$

with its equivalent first-order system

$$\dot{x}_1 = x_2, \qquad \dot{x}_2 = x_2(1 - x_1^2) - x_1. \tag{3.72}$$

In polar coordinates this becomes

$$\begin{aligned} \dot{r} &= r \sin^2\theta(1 - r^2\cos^2\theta) \\ \dot{\theta} &= -1 + \cos\theta \sin\theta(1 - r^2\cos^2\theta). \end{aligned} \tag{3.73}$$

These equations do not give any immediate insight into the nature of the phase portrait which contains a unique attracting limit cycle (see Sections 4.3 and 5.1). In fact, the problem of detecting limit cycles in non-linear systems can be a difficult one which we will now examine more closely.

3.9 Poincaré–Bendixson theory

What are the characteristics of phase portraits with limit cycles? For example, a stable limit cycle requires the existence of a tubular neighbourhood such that all trajectories cross its boundary and tend towards the limit cycle as $t \to \infty$. Suppose there is an annular region, in the phase plane, with the property that all trajectories, that cross its boundary, enter the region. Is this sufficient to ensure that a limit cycle occurs in the annulus? Figure 3.25 shows that the answer to this question is *no*; a saddle and a node can produce trajectories with the

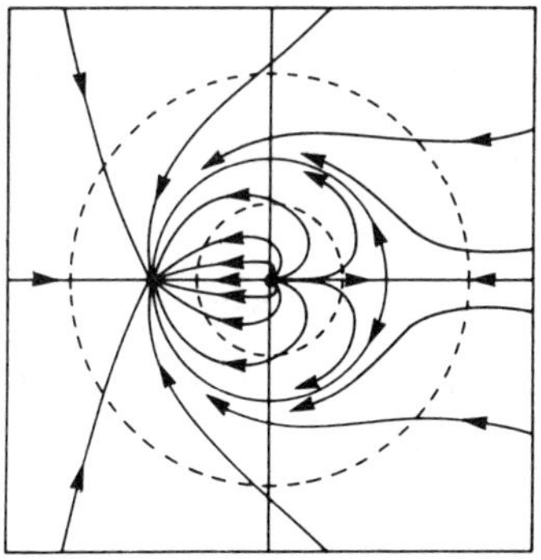

Fig. 3.25. Phase portrait for the system $\dot{r} = r(1-r)$, $\dot{\theta} = \sin\theta$.

properties described. However, the theory to be discussed in this section is a development of this approach. We begin by stating the main theorem without proof (Hirsch and Smale, 1974).

Theorem 3.9.1. Suppose the trajectory $\phi_t(\mathbf{x}_0)$ of the system $\dot{\mathbf{x}} = \mathbf{X}(\mathbf{x})$, with flow ϕ_t, is contained in a bounded region D of the phase plane for $t \geqslant 0$. Then as $t \to \infty$ the points $\phi_t(\mathbf{x}_0)$ must either:

(a) tend to a fixed point; or
(b) spiral towards a limit cycle of the system.

The following property of a phase portrait allows us to obtain a definite result for limit cycles.

Definition 3.9.1. Given the system $\dot{\mathbf{x}} = \mathbf{X}(\mathbf{x})$ with flow ϕ_t; a subset $D \subset \mathsf{R}^2$ is said to be a *positively invariant set* for the system if, for every point $\mathbf{x}_0$ of D, the trajectory $\phi_t(\mathbf{x}_0)$ remains in D for all positive t.

Theorem 3.9.1 then has the following useful corollary.

Corollary. If the region D is a closed, bounded, positively invariant set for the system and there are no fixed points in D, then there must be a limit cycle in D.

Example 3.9.1. Show that the phase portrait of

$$\ddot{x} - \dot{x}(1 - 3x^2 - 2\dot{x}^2) + x = 0$$

has a limit cycle.

Solution. The corresponding first-order system is

$$\dot{x}_1 = x_2, \qquad \dot{x}_2 = -x_1 + x_2(1 - 3x_1^2 - 2x_2^2), \tag{3.74}$$

which becomes

$$\dot{r} = r\sin^2\theta(1 - 3r^2\cos^2\theta - 2r^2\sin^2\theta) \tag{3.75}$$

$$\dot{\theta} = -1 + \tfrac{1}{2}\sin 2\theta(1 - 3r^2\cos^2\theta - 2r^2\sin^2\theta) \tag{3.76}$$

in polar coordinates. Observe:

(a) Equation (3.75) with $r = \frac{1}{2}$ gives

$$\dot{r} = \tfrac{1}{4}\sin^2\theta(1 - \tfrac{1}{2}\cos^2\theta) \geqslant 0 \tag{3.77}$$

with equality only at $\theta = 0$ and π. Thus, $\{\mathbf{x} | r > \frac{1}{2}\}$ is positively invariant;

(b) Equation (3.75) also implies that

$$\dot{r} \leqslant r \sin^2 \theta (1 - 2r^2).$$

Thus for $r = 1/\sqrt{2}$, $\dot{r} \leqslant 0$ with equality only at $\theta = 0$ and π. Thus, $\{\mathbf{x} | r < 1/\sqrt{2}\}$ is positively invariant.

Now (a) and (b) imply that the annular region $\{\mathbf{x} | \frac{1}{2} < r < 1/\sqrt{2}\}$ is positively invariant and, since the only fixed point of (3.74) is at the origin, we conclude there is a limit cycle in the annulus □

The following result gives a condition for there to be *no* limit cycle in a region D.

Theorem 3.9.2. Let D be a simply connected region of the phase plane in which the vector field $\mathbf{X}(\mathbf{x}) = (X_1(x_1, x_2), X_2(x_1, x_2))$ has the property that

$$\frac{\partial X_1}{\partial x_1} + \frac{\partial X_2}{\partial x_2} \tag{3.78}$$

is of constant sign. Then the system $\dot{\mathbf{x}} = \mathbf{X}(\mathbf{x})$ has *no* closed trajectories *wholly* contained in D.

It will be sufficient for our purpose to recognize that a *simply connected region* of the plane is a region with no 'holes' in it (see Fig. 3.26). The theorem follows from Green's theorem in the plane which may be stated as follows:

Let the real valued functions $P(x_1, x_2)$ and $Q(x_1, x_2)$ have continuous first partial derivatives in a simply connected region $\mathcal{R}$ of

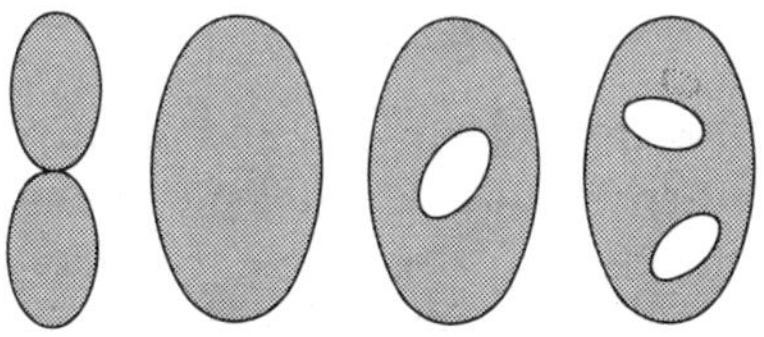

Fig. 3.26. The shaded regions in (a) and (b) have no 'holes' and are simply connected. Those in (c) and (d) have 'holes' and are *not* simply connected.

the x_1x_2-plane bounded by a simple closed curve $\mathscr{C}$. Then

$$\oint_{\mathscr{C}} P\mathrm{d}x_1 + Q\mathrm{d}x_2 = \iint_{\mathscr{R}} \left(\frac{\partial Q}{\partial x_1} - \frac{\partial P}{\partial x_2}\right)\mathrm{d}x_1\mathrm{d}x_2, \tag{3.79}$$

where $\oint_{\mathscr{C}}$ indicates integration along $\mathscr{C}$ in an anticlockwise direction.

To prove Theorem 3.9.2 assume that a limit cycle C of period T exists for the system. Let $P = -X_2$, $Q = X_1$ in (3.79) and obtain

$$\oint_C X_1\mathrm{d}x_2 - X_2\mathrm{d}x_1 = \int_0^T (X_1\dot{x}_2 - X_2\dot{x}_1)\,\mathrm{d}t \qquad (=0) \tag{3.80}$$

$$= \iint_{\mathscr{R}} \left(\frac{\partial X_1}{\partial x_1} + \frac{\partial X_2}{\partial x_2}\right)\mathrm{d}x_1\mathrm{d}x_2 \quad (\neq 0). \tag{3.81}$$

Equation (3.80) follows because C is a solution curve, while (3.81) follows from (3.78). Hence, the closed trajectory C cannot exist.

Example 3.9.2. Prove that if the system

$$\dot{x}_1 = -x_2 + x_1(1 - x_1^2 - x_2^2), \qquad \dot{x}_2 = x_1 + x_2(1 - x_1^2 - x_2^2) + K, \tag{3.82}$$

where K is a constant, has a closed trajectory, then it will either:

(a) encircle the origin; or
(b) intersect the circle $x_1^2 + x_2^2 = \frac{1}{2}$.

Solution. The quantity

$$\frac{\partial X_1}{\partial x_1} + \frac{\partial X_2}{\partial x_2} = 2 - 4(x_1^2 + x_2^2) \tag{3.83}$$

is positive inside the circle $x_1^2 + x_2^2 = \frac{1}{2}$ and negative outside it. Thus, any closed trajectory cannot be wholly contained in the simply connected region $\{(x_1, x_2) \mid x_1^2 + x_2^2 < \frac{1}{2}\}$. Therefore, if a closed trajectory exists it is either contained in $\{(x_1, x_2) \mid x_1^2 + x_2^2 > \frac{1}{2}\}$ or it will intersect $x_1^2 + x_2^2 = \frac{1}{2}$. If the closed orbit is contained in $\{(x_1, x_2) \mid x_1^2 + x_2^2 > \frac{1}{2}\}$ then it must encircle the origin, otherwise it will contain a region of constant negative sign of (3.83). □

Exercises

Sections 3.1–3.3

1. Use the method of isoclines to sketch the global phase portraits of the following systems:
(a) $\dot{x}_1 = x_1 x_2, \quad \dot{x}_2 = \ln x_1, \quad x_1 > 0;$
(b) $\dot{x}_1 = 4x_1(x_2 - 1), \quad \dot{x}_2 = x_2(x_1 + x_1^2);$
(c) $\dot{x}_1 = x_1 x_2, \quad \dot{x}_2 = x_2^2 - x_1^2.$

2. Show that the mapping $(x_1, x_2) \to (f(r)\cos\theta, f(r)\sin\theta)$, where $x_1 = r\cos\theta$, $x_2 = r\sin\theta$ and $f(r) = \tan(\pi r/2r_0)$, is a continuous bijection of $N = \{(x_1, x_2) | r < r_0\}$ onto R^2. Does the bijection map the set of concentric circles on $\mathbf{0}$ in N onto the set of concentric circles on $\mathbf{0}$ in R^2? What property of local phase portraits of linear systems in the plane does this result illustrate?

3. Sketch the local phase portraits of the fixed points in Figs. 3.5(a), 3.21(b) and 3.25.

4. Find the linearizations of the following systems, at the fixed points indicated, by:
  (a) introducing local coordinates at the fixed point;
  (b) using Taylor's theorem.
(i) $\dot{x}_1 = x_1 + x_1 x_2^3/(1 + x_1^2)^2, \quad \dot{x}_2 = 2x_1 - 3x_2, \quad (0, 0);$
(ii) $\dot{x}_1 = x_1^2 + \sin x_2 - 1, \quad \dot{x}_2 = \sinh(x_1 - 1), \quad (1, 0);$
(iii) $\dot{x}_1 = x_1^2 - e^{x_2}, \quad \dot{x}_2 = x_2(1 + x_2), \quad (e^{-1/2}, -1).$
State the preferred method (if one exists) for each system.

5. Use the linearization theorem to classify, where possible, the fixed points of the systems:
(a) $\dot{x}_1 = x_2^2 - 3x_1 + 2, \quad \dot{x}_2 = x_1^2 - x_2^2;$
(b) $\dot{x}_1 = x_2, \quad \dot{x}_2 = -x_1 + x_1^3;$
(c) $\dot{x}_1 = \sin(x_1 + x_2), \quad \dot{x}_2 = x_2;$
(d) $\dot{x}_1 = x_1 - x_2 - e^{x_1}, \quad \dot{x}_2 = x_1 - x_2 - 1;$
(e) $\dot{x}_1 = -x_2 + x_1 + x_1 x_2, \quad \dot{x}_2 = x_1 - x_2 - x_2^2;$
(f) $\dot{x}_1 = x_2, \quad \dot{x}_2 = -(1 + x_1^2 + x_1^4)x_2 - x_1;$
(g) $\dot{x}_1 = -3x_2 + x_1 x_2 - 4, \quad \dot{x}_2 = x_2^2 - x_1^2.$

6. Linearize the system

$$\dot{x}_1 = -x_2, \qquad \dot{x}_2 = x_1 - x_1^5$$

at the origin and classify the fixed point of the *linearized* system. Show that the trajectories of the non-linear system lie on the family of curves

$$x_1^2 + x_2^2 - x_1^6/3 = C,$$

where C is a constant. Sketch these curves to show that the non-linear system and its linearization have qualitatively equivalent local phase portraits at the origin. Why could this conclusion not be deduced from the linearization theorem?

7. Find the principal directions of the fixed points at the origin of the following systems:
(a) $\dot{x}_1 = e^{x_1+x_2} - 1, \quad \dot{x}_2 = x_2;$
(b) $\dot{x}_1 = -\sin x_1 + x_2, \quad \dot{x}_2 = \sin x_2;$
(c) $\dot{x}_1 = \ln(x_1 + x_2 + 1), \quad \dot{x}_2 = \frac{1}{2}x_1 + x_2, \quad x_1 + x_2 > -1;$
and use them to sketch local phase portraits.

Sections 3.4–3.6

8. Find the family of solution curves which satisfy

$$\frac{dx_2}{dx_1} = \frac{x_2^2 - x_1^3}{2x_1x_2}, \qquad x_1, x_2 \neq 0,$$

by making the substitution $x_2^2 = u$. Sketch the family of solutions and hence or otherwise sketch the local phase portrait of the non-simple fixed point of

$$\dot{x}_1 = 2x_1x_2, \qquad \dot{x}_2 = x_2^2 - x_1^3.$$

9. Are the phase portraits of the systems $\dot{x}_1 = x_1 e^{x_1}$, $\dot{x}_2 = x_2 e^{x_1}$ and $\dot{x}_1 = x_1$, $\dot{x}_2 = x_2$ qualitatively equivalent? If so, state the continuous bijection which exhibits the equivalence.

10. Show that the 'straight line' separatrices at the non-simple fixed point of

$$\dot{x}_1 = x_2(3x_1^2 - x_2^2), \qquad \dot{x}_2 = x_1(x_1^2 - 3x_2^2)$$

satisfy $x_2 = kx_1$ where

$$k^2(3 - k^2) = 1 - 3k^2.$$

Hence, or otherwise, find these separatrices and by using isoclines sketch the phase portrait.

11. Show that the system

$$\dot{x}_1 = x_1^2 - x_2^3, \qquad \dot{x}_2 = x_1^2(x_1^2 - x_2^3)$$

has a line of fixed points. Furthermore, show that every fixed point on the line is non-simple. Can this conclusion be reached by using the linearization theorem?

Is the above conclusion true for any system with a line of fixed points?

12. Give an argument to show that a fixed point surrounded by a continuum of closed curves is stable, but not asymptotically stable.

13. Show that the non-linear change of coordinates

$$y_1 = x_1 + x_2^3, \qquad y_2 = x_2 + x_2^2$$

satisfies the requirements of the flow box theorem for the system

$$\dot{x}_1 = -\frac{3x_2^2}{1+2x_2}, \qquad \dot{x}_2 = \frac{1}{1+2x_2}$$

in the neighbourhood of any point (x_1, x_2) with $x_2 \neq -\frac{1}{2}$.

14. Prove that the following systems have no fixed points
(a) $\dot{x}_1 = e^{x_1+x_2}$, $\dot{x}_2 = x_1 + x_2$;
(b) $\dot{x}_1 = x_1 + x_2 + 2$, $\dot{x}_2 = x_1 + x_2 + 1$;
(c) $\dot{x}_1 = x_2 + 2x_2^3$, $\dot{x}_2 = 1 + x_2^2$
and sketch their phase portraits.

15. Sketch phase portraits consistent with the following information:
(a) two fixed points, a saddle and a stable node;
(b) three fixed points, one saddle and two stable nodes.

16. Find the local phase portraits at each of the fixed points of the system

$$\dot{x}_1 = x_1(1 - x_1^2), \qquad \dot{x}_2 = x_2.$$

Use these results to suggest a global phase portrait. Check whether your suggestion is correct by using the method of isoclines.

Section 3.7

17. Find first integrals of the following systems together with their domains of definition.
(a) $\dot{x}_1 = x_2, \quad \dot{x}_2 = x_1^2 + 1;$
(b) $\dot{x}_1 = x_1(x_2 + 1), \quad \dot{x}_2 = -x_2(x_1 + 1);$
(c) $\dot{x}_1 = \sec x_1, \quad \dot{x}_2 = -x_2^2, \quad |x_1| < \pi/2;$
(d) $\dot{x}_1 = x_1(x_1 e^{x_2} - \cos x_2), \quad \dot{x}_2 = \sin x_2 - 2x_1 e^{x_2}.$

18. Find a first integral of the system

$$\dot{x}_1 = x_1 x_2 - 3x_1^3, \qquad \dot{x}_2 = x_2^2 - 6x_1^2 x_2 + x_1^4$$

using the substitution $x_2 = ux_1^2$. Sketch the phase portrait.

19. How do the phase portraits of the two systems

$$\dot{x}_1 = x_1(x_2^2 - x_1), \qquad \dot{x}_2 = -x_2(x_2^2 - x_1)$$
$$\dot{x}_1 = x_1, \qquad \dot{x}_2 = -x_2$$

differ?

20. Find a first integral of the system

$$\dot{x}_1 = x_1 x_2, \qquad \dot{x}_2 = \ln x_1, \qquad x_1 > 1$$

in the region indicated. Hence, or otherwise, sketch the phase portrait.

21. Find first integrals for the linear systems $\dot{\boldsymbol{x}} = \boldsymbol{J}\boldsymbol{x}$, where $\boldsymbol{J} = 2 \times 2$ Jordan matrix of node, centre and focus type. State maximal regions on which these first integrals exist.

Is a system which has an asymptotically stable fixed point ever conservative?

22. Find a Hamiltonian H for a particle moving along a straight line subject to

$$\ddot{x} = -x + \alpha x^2,$$

$\alpha > 0$, where x is the displacement. Sketch the level curves of the

Hamiltonian H in the phase plane. Indicate the regions of the phase plane that contain trajectories which give non-oscillatory motion.

Sections 3.8–3.9

23. Sketch phase portraits consistent with the following information:

(a) an unstable limit cycle and three fixed points, one saddle and two stable nodes;

(b) a stable focus and two limit cycles, one stable and one unstable.

24. For each of the following systems show that the indicated region R is positively invariant:

(a) $\dot{x}_1 = 2x_1x_2, \quad \dot{x}_2 = x_2^2, \quad R = \{(x_1,x_2)|x_2 \geqslant 0\}$;

(b) $\dot{x}_1 = -\alpha x_1 + x_2, \quad \dot{x}_2 = (\beta - \alpha)x_2; \quad \alpha, \beta$ constant, $R = \{(x_1,x_2)|x_2 = \beta x_1\}$;

(c) $\dot{x}_1 = -x_1 + x_2 + x_1(x_1^2 + x_2^2), \quad \dot{x}_2 = -x_1 - x_2 + x_2(x_1^2 + x_2^2)$, $R = \{(x_1,x_2)|x_1^2 + x_2^2 < 1\}$;

(d) $\dot{x}_1 = x_1(x_2^2 - x_1), \dot{x}_2 = -x_2(x_2^2 - x_1), \quad R = \{(x_1,x_2)|x_1 > x_2^2\}$.

25. Find closed trajectories of the following differential equations:

(a) $\ddot{x} + (2\dot{x}^2 + x^4 - 1)\dot{x} + x^3 = 0$; (b) $\ddot{x} + (\dot{x}^2 + x^2 - 4)x = 0$;

(c) $\ddot{x} + (2\dot{x}^2 + x^4 - 1)x^3 = 0$.

26. Show that the polar form of the non-linear system

$$\dot{x}_1 = -x_2 + x_1(1 - x_1^2 - x_2^2), \qquad \dot{x}_2 = x_1 + x_2(1 - x_1^2 - x_2^2)$$

is given by

$$\dot{r} = r(1 - r^2), \qquad \dot{\theta} = 1.$$

Solve this equation subject to the initial conditions $r(0) = r_0, \theta(0) = \theta_0$ at $t = 0$ to obtain

$$r(t) = r_0/(r_0^2 + (1 - r_0^2)e^{-2t})^{1/2}.$$

Plot the graph of $r(t)$ against t for

(a) $0 < r_0 < 1$, (b) $r_0 = 1$, (c) $r_0 > 1$,

and obtain the phase portrait of the system. Can the phase portrait be sketched more easily from the polar differential equation?

27. Prove that there exists a region $R = \{(x_1, x_2) | x_1^2 + x_2^2 \leqslant r^2\}$ such that all trajectories of the system

$$\dot{x}_1 = -wx_2 + x_1(1 - x_1^2 - x_2^2), \qquad \dot{x}_2 = wx_1 + x_2(1 - x_1^2 - x_2^2) - F,$$

where w and F are constants, eventually enter R. Show that the system has a limit cycle when $F = 0$.

28. Prove that the system

$$\dot{x}_1 = 1 - x_1 x_2, \qquad \dot{x}_2 = x_1$$

has no limit cycles.

29. Consider the system

$$\begin{aligned} \dot{x}_1 &= -wx_2 + x_1(1 - x_1^2 - x_2^2) - x_2(x_1^2 + x_2^2), \\ \dot{x}_2 &= wx_1 + x_2(1 - x_1^2 - x_2^2) + x_1(x_1^2 + x_2^2) - F, \end{aligned}$$

where w and F are constants. Prove that if the system has a limit cycle such that all of its points are at a distance greater than $1/\sqrt{2}$ from the origin, then the limit cycle must encircle the origin.

30. Suppose that the region $R = \{(x_1, x_2) | x_1, x_2 > 0\}$ is positively invariant for the system $\dot{x}_1 = X_1(x_1, x_2), \dot{x}_2 = X_2(x_1, x_2)$ and that

$$\begin{aligned} \dot{x}_1 &\lesseqgtr 0 \quad \text{for } x_2 \gtreqless -x_1^2 + 3x_1 + 1 \\ \dot{x}_2 &\lesseqgtr 0 \quad \text{for } x_2 \gtreqless x_1, \end{aligned}$$

respectively. Assuming that there are no closed orbits in R prove that the unique fixed point in R is asymptotically stable.

CHAPTER FOUR

Applications

The results of Chapters 2 and 3 play an important role in the construction and analysis of models of real time-dependent systems. In this chapter, we illustrate these applications by presenting a number of models from several different areas of study.

In each model the vector $\mathbf{x}(t)$ gives the *state* of the system at time t (cf. Section 1.2.2). The time-development of the states is governed by the *dynamical equations* (or equations of motion) $\dot{\mathbf{x}} = \mathbf{X}(\mathbf{x})$ and the qualitative behaviour of the evolution of the states is given by the corresponding phase portrait.

4.1 Linear models

A model is said to be linear if it has linear dynamical equations. As we have seen in Chapter 2, such equations only show certain kinds of qualitative behaviour. For example, in the plane the qualitative behaviour is confined to the classification given in Section 2.4.

4.1.1 *A mechanical oscillator*

Consider a mass m supported on a vertically mounted spring as shown in Fig. 4.1. The mass is constrained to move only along the axis of the spring. The mass is also attached to a piston that moves in a cylinder of fluid contained in the spring. The piston resists any motion of the

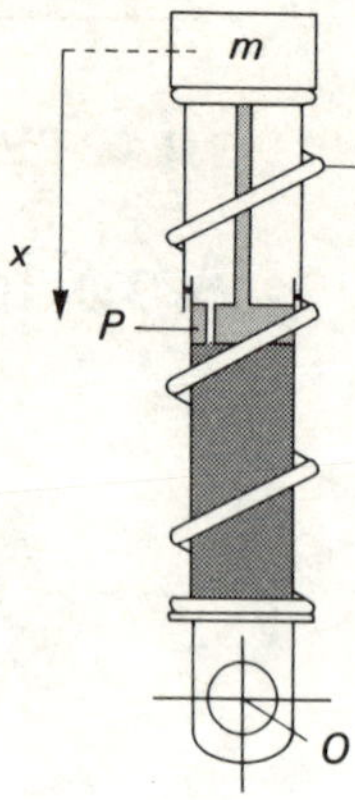

Fig. 4.1. Mass m supported on vertically mounted coiled spring S rigidly fixed at its lower end O. The mass is also attached to the piston P. P moves in a fluid filled cylinder and this impedes the motion of m.

mass. This arrangement is an idealization of the shock absorbers that can be seen on most motorcycles.

Let x be the displacement of the mass m below its equilibrium position at rest on the spring. Assume that:

(a) the spring obeys Hooke's law so that it exerts a restoring force of Kx ($K > 0$) on the mass;
(b) the force exerted by the piston, which opposes the motion of the mass, is $2k$ ($k \geqslant 0$) times the momentum p of the mass.

With these assumptions, the dynamical equations which model the motion of the mass are linear. They can be constructed by observing that:

(a) the momentum p of the mass is given by $m\dot{x} = p$;
(b) the rate of change of linear momentum of the mass is equal to the force applied to it (i.e. Newton's second law of motion).

The second observation gives

$$\dot{p} = -K(l+x) - 2kp + mg \tag{4.1}$$

where l is the compression of the spring at equilibrium. However, at equilibrium $\dot{p} = p = x = 0$, so $Kl = mg$ and (4.1) can be written as

$$\dot{p} = m\ddot{x} = -Kx - 2km\dot{x}.$$

Thus, the position of the mass satisfies the linear second-order equation

$$\ddot{x} + 2k\dot{x} + \omega_0^2 x = 0, \tag{4.2}$$

where $\omega_0^2 = K/m > 0$ and $k \geqslant 0$. An equivalent first-order system is given by putting $x_1 = x$ and $x_2 = \dot{x}$ (cf. Exercise 1.28) to obtain

$$\dot{x}_1 = x_2; \qquad \dot{x}_2 = -\omega_0^2 x_1 - 2kx_2. \tag{4.3}$$

The linear system (4.3) has coefficient matrix

$$\boldsymbol{A} = \begin{bmatrix} 0 & 1 \\ -\omega_0^2 & -2k \end{bmatrix}, \tag{4.4}$$

with $\mathrm{Tr}(\boldsymbol{A}) = -2k \leqslant 0$ and $\det(\boldsymbol{A}) = \omega_0^2 > 0$. It follows that the fixed point at the origin of the phase portrait of (4.3) is always stable (asymptotically for $k > 0$). As Fig. 4.2 shows, for each fixed value of ω_0^2 (i.e. the spring constant K), the phase portrait passes through the sequence: centre, foci, improper nodes, nodes; as k increases through the interval $0 \leqslant k < \infty$.

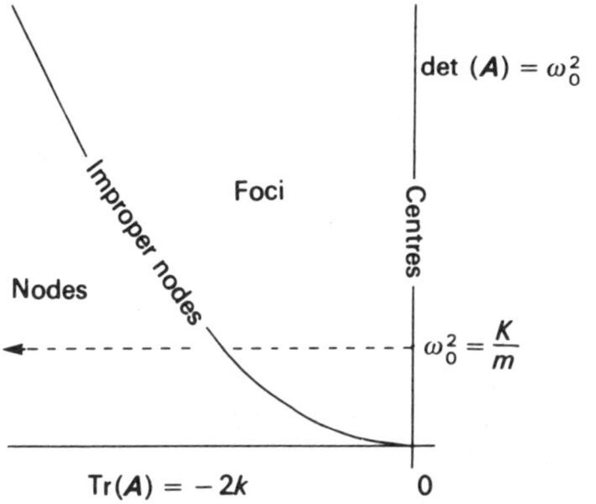

Fig. 4.2. Quadrant of the $\mathrm{Tr}(\boldsymbol{A}) - \det(\boldsymbol{A})$ plane reached by $\boldsymbol{A}$ in (4.4). $\mathrm{Tr}(\boldsymbol{A}) \leqslant 0$ implies that all phase portraits are stable (cf. Fig. 2.7).

Each of the above types of phase portrait corresponds to a qualitatively different motion of the mass m. There are four cases:

(*a*) $k = 0$

$\mathrm{Tr}(\boldsymbol{A}) = 0$ and the eigenvalues of $\boldsymbol{A}$ are $\lambda_1 = -\lambda_2 = i\omega_0$. The corresponding canonical system $\dot{\boldsymbol{y}} = \boldsymbol{J}\boldsymbol{y}$ has solution (2.53) with $\beta = \omega_0$ and $\dot{\boldsymbol{x}} = \boldsymbol{A}\boldsymbol{x}$ has trajectories

$$(x_1(t), x_2(t)) = (R\cos(\omega_0 t + \theta), -\omega_0 R\sin(\omega_0 t + \theta)), \qquad (4.5)$$

(see Exercise 4.1). Notice that for this problem $x_2(t) = \dot{x}_1(t)$. The phase portrait is therefore as shown in Fig. 4.3. The relationship between the oscillations in x_1 and the trajectories is also illustrated in this figure.

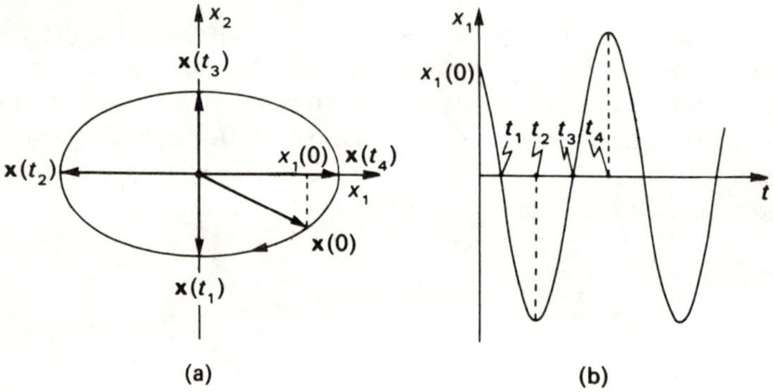

Fig. 4.3. (a) Phase portrait for $\dot{x}_1 = x_2$, $\dot{x}_2 = -\omega_0^2 x_1$ consists of a continuum of concentric ellipses. The trajectory (4.5) through $\mathbf{x}(0) = (R\cos\theta, -\omega_0 R\sin\theta)$ corresponds to the oscillations in x_1 shown in (b).

The motion of the mass consequently consists of *persistent* oscillations of both position and velocity with the same period $T_0 = 2\pi/\omega_0$. The mass is said to execute *free* (no external applied forces), *undamped* ($k = 0$) oscillations at the *natural frequency* $\omega_0 = K/m$ of the system.

(*b*) $0 < k < \omega_0$

The eigenvalues of $\boldsymbol{A}$ are now $-k \pm i(\omega_0^2 - k^2)^{1/2}$ and (2.52) implies that

$$x_1(t) = Re^{-kt}\cos(\beta t + \theta) \qquad (4.6)$$

where $\beta = (\omega_0^2 - k^2)^{1/2}$ (see Exercise 4.1).

The motion is, therefore, modified in two ways:

(a) the amplitude of the oscillation decays with increasing t; and
(b) the period of the oscillations is $T = 2\pi/\beta > T_0$.

The system is said to undergo *damped*, free oscillations. These oscillations, and the corresponding phase portrait, are shown in Fig. 4.4.

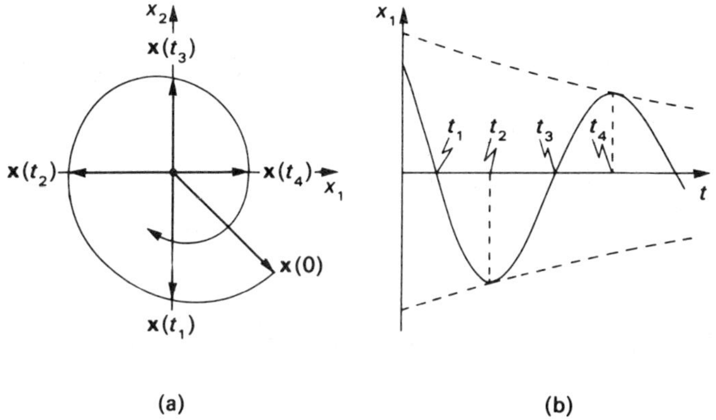

Fig. 4.4. Phase portrait, (a), and $x_1(t)$ versus t, (b), for $\dot{x}_1 = x_2, \dot{x}_2 = -\omega_0^2 x_1 - 2kx_2$, $0 < k < \omega_0$. The phase portrait is a stable focus and the envelope of the oscillations – shown dashed in (b) – is Re^{-kt}.

(*c*) $k = \omega_0$

As $k \to \omega_{0-}$ the period $T = 2\pi/(\omega_0^2 - k^2)^{1/2}$ increases indefinitely and finally at $k = \omega_0$ the oscillations disappear. The eigenvalues of $\boldsymbol{A}$ are real and both equal to $-k$. We have reached the line of stable improper nodes in Fig. 4.2. The system is said to be *critically damped*, in so much as the oscillations of the *underdamped* ($0 < k < \omega_0$) system have just disappeared.

In this case we have from (2.48)

$$\mathbf{x}(t) = e^{-kt}(a + bt, b - k(a + bt)) \tag{4.7}$$

with $a, b \in \mathrm{R}$ (see Exercise 4.1). The motion of the mass depends on the initial conditions as the phase portrait in Fig. 4.5(a) shows.

Suppose the mass is projected with velocity $x_2 = -s_0$ (speed $s_0 > 0$) from $x_1 = x_0 > 0$. The phase portrait shows that if

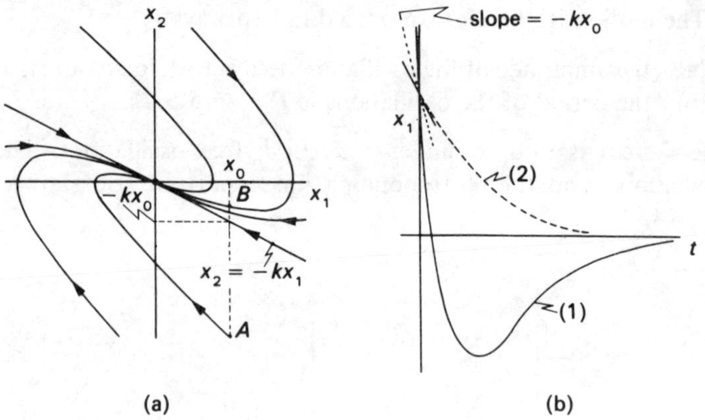

Fig. 4.5. (a) Phase portrait for $\dot{x}_1 = x_2$, $\dot{x}_2 = -k^2x_1 - 2kx_2$: the origin is an improper node. The trajectory through A gives $x_1(t)$ shown by curve (1) in (b), while the trajectory through B gives that shown in curve (2).

$s_0 > kx_0$ (e.g. point A in Fig. 4.5(a)) then the mass overshoots the equilibrium position once and then approaches $x_1 = 0$. If $s_0 < kx_0$ (e.g. point B) then no overshoot occurs. The two kinds of behaviour of x_1 are illustrated in Fig. 4.5(b).

(*d*) $k > \omega_0$

The eigenvalues of $\boldsymbol{A}$ are no longer equal, but both are negative (i.e. $\lambda_1 = -k + (k^2 - \omega_0^2)^{1/2}$, $\lambda_2 = -k - (k^2 - \omega_0^2)^{1/2}$) and so the origin is a stable node. It can be shown that

$$x_1(t) = a\mathrm{e}^{\lambda_1 t} + b\mathrm{e}^{\lambda_2 t}, \qquad x_2(t) = \dot{x}_1(t) \tag{4.8}$$

and the phase portraits look like those in Fig. 4.6.

As $k \to \infty$, $\lambda_1 \to 0$ and $\lambda_2 \to -\infty$ so the principal directions at the fixed point move to coincide with the coordinate axes as indicated in Fig. 4.6. The trajectories steepen and the speed, $|x_2|$, decreases rapidly from large values while the mass moves through a relatively short distance. The area of the plane from which overshoot can occur diminishes and the system is said to be *overdamped*.

The different motions of the mass described in (a)–(d) are easily recognizable in the context of vehicle suspensions. The overdamped case would be a 'very hard' suspension transmitting shocks almost

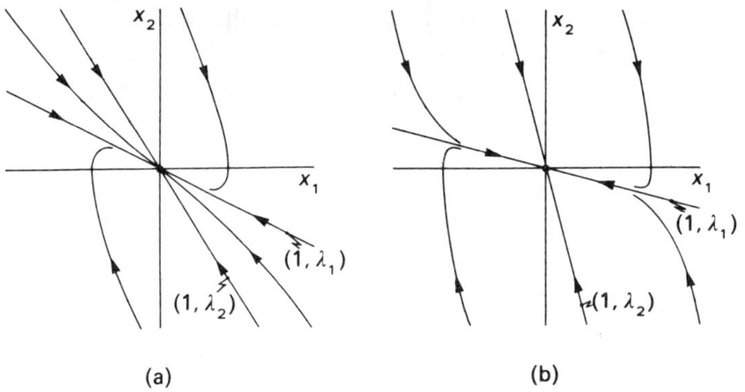

Fig. 4.6. Phase portraits for $\dot{x}_1 = x_2, \dot{x}_2 = -\omega_0^2 x_1 - 2kx_2$ with (a) $k = k_1 > \omega_0$; (b) $k = k_2 > k_1 > \omega_0$. The principal directions are given by $(1, \lambda_1)$, $(1, \lambda_2)$ (see Exercise 4.2).

directly to the vehicle. The highly underdamped case would allow the vehicle to 'wallow'. These situations would both be unsatisfactory. The critically damped case, with at most one overshoot, would obviously be most reasonable. Indeed, this is usually the case as the reader may verify on the suspension of any motor cycle or car.

The equations (4.2) or (4.3) are arguably the most frequently used linear dynamical equations. They appear in models of all kinds of time dependent systems (see Sections 4.1.2 and 4.1.3). Their solutions are characterized by trigonometric (or harmonic) oscillations, which are damped when $k > 0$. As a result (4.2) and (4.3) are known as the *damped harmonic oscillator* equations.

4.1.2 *Electrical circuits*

Electrical circuit theory is a rich source of both linear and non-linear dynamical equations. We will briefly review the background for readers who are not familiar with the subject.

An electrical circuit is a collection of 'circuit elements' connected in a network of closed loops. The notation for some typical circuit elements is given in Fig. 4.7, along with the units in which they are measured.

Differences in *electrical potential* (measured in volts) cause *charge* to move through a circuit. This flow of charge is called a *current*, j (measured in amperes). We can think of an electrical circuit as a set of

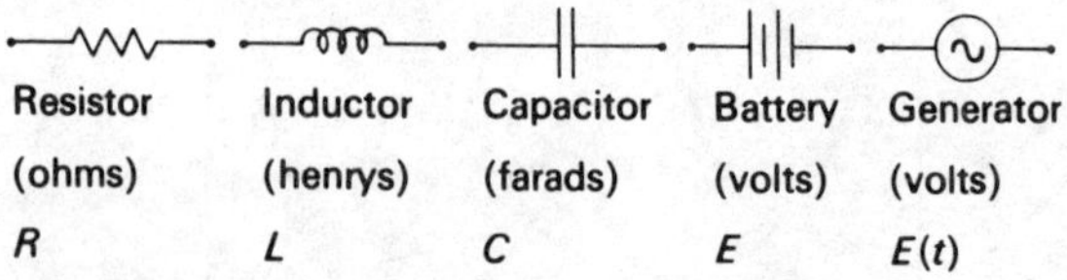

Fig. 4.7. The values R, L and C of the resistor, inductor and capacitor are always non-negative (unless otherwise stated) and independent of t. Batteries and generators are sources of electrical potential difference.

'nodes' or 'terminals' with circuit elements connected between them. If a circuit element is connected between node n and node m of a circuit we associate with it:

(a) a potential difference or *voltage* v_{nm};
(b) a current j_{nm}.

The voltage v_{nm} is the potential difference between node n and node m, so that $v_{nm} = -v_{mn}$. Equally j_{nm} measures the flow of charge from n to m and therefore $j_{nm} = -j_{mn}$.

The currents and voltages associated with the elements in a circuit are related in a number of ways. The potential differences satisfy *Kirchhoff's voltage law*:

> The sum of the potential differences around any closed loop in a circuit is zero; (4.9)

and the currents satisfy *Kirchhoff's current law*:

> The sum of the currents flowing *into* a node is equal to the sum of the currents flowing *out* of it. (4.10)

Apart from these fundamental laws, the current flowing through a resistor, inductor or capacitor is related to the voltage across it. If a resistor is connected between nodes n and m then

$$v_{nm} = j_{nm}R. \tag{4.11}$$

This is *Ohm's law* and such a resistor is said to be *ohmic*. More generally, the relationship is non-linear with

$$v_{nm} = f(j_{nm}). \tag{4.12}$$

Unless otherwise stated we will assume (4.11) to be valid.

The inductor and capacitor provide relationships involving time derivatives which, in turn, lead to dynamic equations. With the notation defined in Fig. 4.8, these relations are:

$$v = L\frac{\mathrm{d}j}{\mathrm{d}t}; \tag{4.13}$$

and

$$C\frac{\mathrm{d}v}{\mathrm{d}t} = j. \tag{4.14}$$

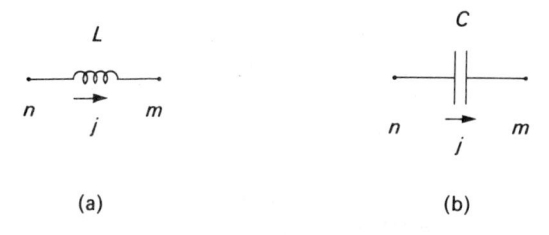

Fig. 4.8. In (4.13) and (4.14), $j \equiv j_{nm}$ and $v \equiv v_{nm}$; the potential difference is taken in the direction of j.

Example 4.1.1. Find the dynamical equations of the LCR circuit shown in Fig. 4.9. Show that, with $x_1 = v_{23}$ and $x_2 = \dot{v}_{23}$, these equations can be written in the form (4.3).

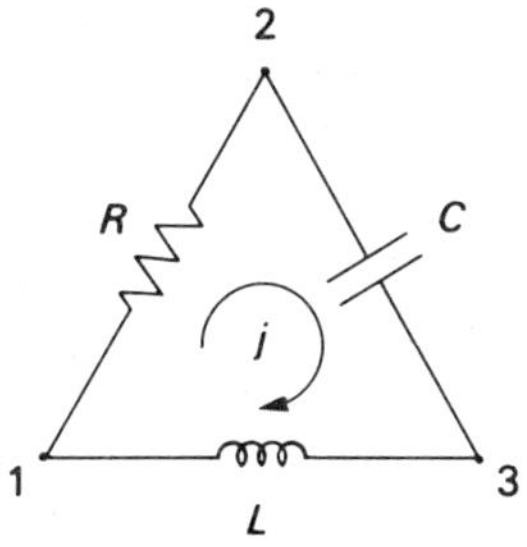

Fig. 4.9. The 'series' LCR circuit.

Solution. Kirchhoff's current law is already satisfied by assigning a current, j, to the single closed loop as shown; while the voltage law implies

$$v_{12} + v_{23} + v_{31} = 0. \tag{4.15}$$

The circuit element relations give

$$v_{12} = jR, \tag{4.16}$$

$$v_{31} = L\frac{\mathrm{d}j}{\mathrm{d}t} \tag{4.17}$$

and

$$C\frac{\mathrm{d}v_{23}}{\mathrm{d}t} = j. \tag{4.18}$$

Substituting (4.15) and (4.16) into (4.17) and writing $v_{23} \equiv v$ gives

$$\frac{\mathrm{d}v}{\mathrm{d}t} = \frac{j}{C}, \qquad \frac{\mathrm{d}j}{\mathrm{d}t} = -\frac{R}{L}j - \frac{v}{L}. \tag{4.19}$$

Let $x_1 = v$ and $x_2 = \dot{v} = j/C$ and (4.19) becomes

$$\dot{x}_1 = x_2, \qquad \dot{x}_2 = -\omega_0^2 x_1 - 2kx_2. \tag{4.20}$$

with $\omega_0^2 = 1/LC > 0$ and $2k = R/L \geqslant 0$. Equation (4.20) is precisely the same system as (4.3) for the mechanical oscillator. □

The fact that (4.20) and (4.3) are the same means that the series LCR circuit is an electrical analogue of the mechanical oscillator. The potential difference v ($= x_1$) across the capacitor behaves in exactly the same way with time as the displacement of the mass on the spring. Clearly, (4.20) implies

$$\ddot{v} + 2k\dot{v} + \omega_0^2 v = 0 \tag{4.21}$$

which is exactly the same as (4.2).

The phase portraits in Fig. 4.3–4.6 can be re-interpreted in terms of $x_1 = v$ and $x_2 = j/C$. For given L and C, the capacitor voltage may: oscillate without damping ($R = 0$); execute damped oscillations $[0 < R < 2(L/C)^{1/2}]$; be critically damped $[R = 2(L/C)^{1/2}]$ or over-damped $[R > 2(L/C)^{1/2}]$.

Observe that the sign of j determines the sense in which charge flows around the circuit. If $j > 0$, the flow is clockwise; if $j < 0$ the flow is counterclockwise (cf. Fig. 4.9). Consider the case when $R = 0$, for example, and suppose $v(0) = v_0 > 0, j(0) = 0$. Equation (4.5) implies that

$$\left(v(t), \frac{j(t)}{C}\right) = \left(v_0 \cos\left(\frac{t}{\sqrt{(LC)}}\right), -\frac{v_0}{\sqrt{(LC)}} \sin\left(\frac{t}{\sqrt{(LC)}}\right)\right). \quad (4.22)$$

Charge flows around the circuit in a *counter-clockwise* sense until $t = \pi\sqrt{(LC)}$, when

$$v(\pi\sqrt{(LC)}) = -v_0 \quad \text{and} \quad j(\pi\sqrt{(LC)}) = 0. \quad (4.23)$$

As t increases further, $v(t)$ increases and $j(t)$ becomes positive. Charge now flows around the circuit in a *clockwise* sense until $t = 2\pi\sqrt{(LC)}$ when

$$v(2\pi\sqrt{(LC)}) = v_0 \quad \text{and} \quad j(2\pi\sqrt{(LC)}) = 0. \quad (4.24)$$

The state of the circuit is now the same as it was at $t = 0$ and the oscillations continue. This behaviour is easily recognized in the elliptical trajectories of Fig. 4.3(a).

4.1.3 *Economics*

An economy is said to be *closed* if all its output is either consumed or invested within the economy itself. There are no exports, imports or external injections of capital. Thus, if Y, C and I are, respectively, the output, consumption and investment for a closed economy at time t, then $Y = C + I$. When external capital injections, such as government expenditure G, are included the economy is no longer closed and output is increased so that

$$Y = C + I + G. \quad (4.25)$$

Furthermore, consumption increases with output; we take

$$C = dY = (1 - s)Y, \quad (4.26)$$

where $d, s > 0$ are, respectively, the marginal propensities to consume and to save (see Hayes, 1975).

Let us consider an economy which operates with a constant level of government spending G_0. At any particular time t, the economy is subject to a demand $D(t)$ which measures the desired level of consumption and investment. The aim is to balance the economy so that output matches demand, i.e. $D(t) \equiv Y(t)$. However, in practice output is unable to respond instantaneously to demand. There is a response time, τ, associated with development of new plant, etc.

To achieve balance in the presence of lags we must plan ahead and adjust output to meet predicted demand, by taking

$$D(t) = (1-s)Y(t-\tau) + I(t) + G_0. \tag{4.27}$$

In (4.27) we have assumed that investment does not change significantly during the period τ so that $I(t-\tau) = I(t)$. If we now take

$$Y(t-\tau) = Y(t) - \tau\dot{Y}(t) + 0(\tau^2), \tag{4.28}$$

we see that (4.27) means that balance is achieved to first order in τ provided

$$(1-s)\tau\dot{Y}(t) = -sY(t) + I(t) + G_0, \tag{4.29}$$

for every real t.

Although $I(t)$ does not change significantly in periods of order τ; it is not a constant. Investment will respond to trends in output. One plausible investment policy – the 'accelerator principle' – claims that $I(t) \equiv a\dot{Y}(t)$, $a > 0$, is suitable. Time lags in implementing investment decisions prevent this relation from being satisfied, however, we always move towards this ideal if we take

$$\dot{I}(t) = b(a\dot{Y}(t) - I(t)), \qquad b > 0. \tag{4.30}$$

Equations (4.29) and (4.30) are the basis of the dynamics of the economy. They can be cast into more familiar form by differentiating (4.29) to obtain

$$(1-s)\tau\ddot{Y} + s\dot{Y} = \dot{I} = b(a\dot{Y} - I). \tag{4.31}$$

Substituting for I from (4.29) finally gives

$$(1-s)\tau\ddot{y} + (s - ba + (1-s)\tau b)\dot{y} + sby = 0, \tag{4.32}$$

where $y = Y - (G_0/s)$. Thus the output y, relative to G_0/s, satisfies (4.2) with

$$k = (s - ba + (1-s)\tau b)/2\tau(1-s) \text{ and } \omega_0^2 = sb/\tau(1-s) > 0. \tag{4.33}$$

When k is non-negative, the phase plane analysis of Section 4.1.1 applies, with suitable re-interpretation, to the output of the economy. In particular, the oscillations in Y (about G_0/s) correspond to the 'booms' and 'depressions' experienced by many economies.

A new feature in this model is that the 'damping constant', k, given in (4.33) can be *negative*. This means that the equivalent first-order system

$$\dot{x}_1 = x_2, \qquad \dot{x}_2 = -\omega_0^2 x_1 + 2|k|x_2 \tag{4.34}$$

has an unstable fixed point at the origin of its phase plane. It has the qualitative behaviour characteristic of the positive quadrant of the Tr($\boldsymbol{A}$)–det($\boldsymbol{A}$) plane given in Fig. 2.7. Under such circumstances, the peaks of the booms increase with t; as do the depths of the depressions. External intervention is then necessary to prevent output from eventually reaching zero.

4.1.4 *Coupled oscillators*

The decoupling of linear systems with dimension > 2 into subsystems of lower dimension (see Section 2.7) is the basis of the treatment of several models involving interacting oscillators. Small oscillations are usually assumed in order to ensure that the dynamical equations are linear.

Consider two identical simple pendula connected by a light spring as illustrated in Fig. 4.10. We will examine the motion of this system in the vertical plane through OO'.

The dynamical equations can be obtained from the components of force perpendicular to the pendula rods. For small oscillations we obtain

$$\begin{aligned} \ddot{\theta} &= -\frac{g}{a}\theta - \frac{\lambda l}{a}(\theta - \phi) \\ \ddot{\phi} &= -\frac{g}{a}\phi + \frac{\lambda l}{a}(\theta - \phi). \end{aligned} \tag{4.35}$$

If the unit of time is chosen so that $t\sqrt{(g/a)}$ is replaced by t and we take

$$x_1 = \theta, \qquad x_2 = \dot{\theta}, \qquad x_3 = \phi \text{ and } x_4 = \dot{\phi}, \tag{4.36}$$

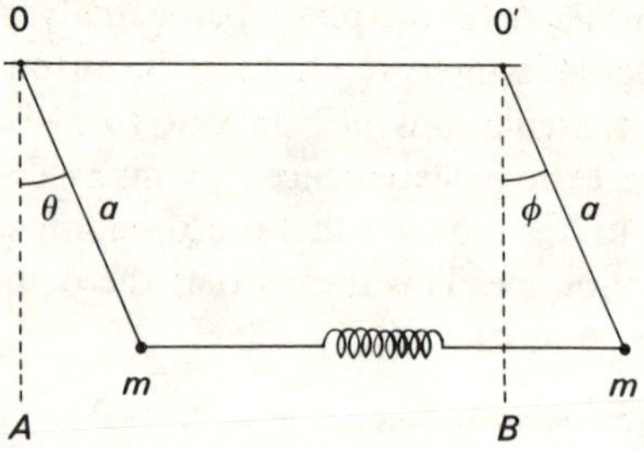

Fig. 4.10. Two identical pendula, A and B, of length a are suspended at points O and O' and connected by a spring of natural length $l = OO'$ and spring constant $m\lambda$. The angles of deflection from the vertical are θ and ϕ as shown.

(4.35) can be written as the first-order linear system,

$$\begin{aligned} \dot{x}_1 &= x_2, & \dot{x}_2 &= -x_1 - \alpha(x_1 - x_3), \\ \dot{x}_3 &= x_4, & \dot{x}_4 &= -x_3 + \alpha(x_1 - x_3), \end{aligned} \tag{4.37}$$

where $\alpha = \lambda l/g > 0$.

The system (4.37) can be decoupled into two-dimensional subsystems by the linear change of variable

$$\begin{aligned} x_1' &= x_1 + x_3, & x_2' &= x_2 + x_4, \\ x_3' &= x_1 - x_3, & x_4' &= x_2 - x_4. \end{aligned} \tag{4.38}$$

The transformed system has coefficient matrix

$$\left[\begin{array}{cc|cc} 0 & 1 & 0 & 0 \\ -1 & 0 & 0 & 0 \\ \hline 0 & 0 & 0 & 1 \\ 0 & 0 & -(1+2\alpha) & 0 \end{array}\right]. \tag{4.39}$$

The diagonal blocks correspond to a pair of undamped harmonic oscillators, one with $\omega_0 = 1$ and the other with $\omega_0 = \sqrt{(1+2\alpha)}$. Of course, (4.39) is not a Jordan form; a further change

$$\begin{aligned} x_1' &= y_1, & x_2' &= -y_2, \\ x_3' &= y_3, & x_4' &= -\sqrt{(1+2\alpha)}y_4, \end{aligned} \tag{4.40}$$

is necessary to obtain the canonical system $\dot{\boldsymbol{y}} = \boldsymbol{J}\boldsymbol{y}$ with

$$\boldsymbol{J} = \left[\begin{array}{cc|cc} 0 & -1 & 0 & 0 \\ 1 & 0 & 0 & 0 \\ \hline 0 & 0 & 0 & -\sqrt{(1+2\alpha)} \\ 0 & 0 & \sqrt{(1+2\alpha)} & 0 \end{array}\right]. \tag{4.41}$$

The solutions of (4.37) are linear combinations of those of $\dot{\boldsymbol{y}} = \boldsymbol{J}\boldsymbol{y}$ and a number of interesting features appear for different initial conditions. For example:

(a) $\theta = \phi = 0,\ \dot{\theta} = \dot{\phi} = v$

Equations (4.36), (4.38) and (4.40) imply that $y_1 = y_3 = y_4 = 0$ and $y_2 = -2v$ at $t = 0$; therefore (cf. (2.53)),

$$y_1 = 2v\sin(t), \qquad y_2 = -2v\cos(t), \qquad y_3 \equiv y_4 \equiv 0. \tag{4.42}$$

Transformation back to the natural variables gives

$$\theta = v\sin(t), \qquad \phi = v\sin(t). \tag{4.43}$$

Thus $\theta \equiv \phi$ and the pendula oscillate in phase with period 2π; the spring remains unstretched at all times;

(b) $\theta = \phi = 0,\ \dot{\theta} = -\dot{\phi} = v$

In this case, we find

$$\begin{aligned} y_1 \equiv y_2 \equiv 0, \qquad y_3 &= \frac{2v}{\sqrt{(1+2\alpha)}}\sin(\sqrt{(1+2\alpha)}t) \\ y_4 &= \frac{-2v}{\sqrt{(1+2\alpha)}}\cos(\sqrt{(1+2\alpha)}t) \end{aligned} \tag{4.44}$$

and

$$\theta = -\phi = \frac{v}{\sqrt{(1+2\alpha)}}\sin(\sqrt{(1+2\alpha)}t) \tag{4.45}$$

so that the pendula oscillate with period $2\pi/\sqrt{(1+2\alpha)}$ and a phase difference of π. They are always symmetrically placed relative to the vertical bisecting OO'.

These two special forms of oscillation are called the *normal modes* of the coupled pendula. They correspond to the solutions of the canonical system (4.41) in which oscillations take place in one of the subsystems while the other has solutions which are identically zero. In (a) one *normal coordinate* $y_1 = \theta + \phi$ oscillates while the other $y_3 = \theta - \phi \equiv 0$. In (b) these roles are reversed and $y_1 \equiv 0$.

Another kind of motion is obtained if the initial conditions are $\theta = \phi = \dot{\theta} = 0$ and $\dot{\phi} = v$ when $t = 0$, so that $y_1(0) = y_3(0) = 0$, $y_2(0) = -v$ and $y_4(0) = v/\sqrt{(1+2\alpha)}$. In this case,

$$y_1 = v \sin(t) \quad \text{and} \quad y_3 = \frac{-v}{\sqrt{(1+2\alpha)}} \sin(\sqrt{(1+2\alpha)}t) \tag{4.46}$$

so that

$$\begin{aligned} \theta &= \frac{v}{2}\left\{\sin(t) - \frac{\sin(\sqrt{(1+2\alpha)}t)}{\sqrt{(1+2\alpha)}}\right\} \\ \phi &= \frac{v}{2}\left\{\sin(t) + \frac{\sin(\sqrt{(1+2\alpha)}t)}{\sqrt{(1+2\alpha)}}\right\} \end{aligned} \tag{4.47}$$

Suppose, now, that the spring constant λ is sufficiently small for $0 < \alpha \ll 1$, so that the coupling is light. Since

$$\sin(t) \pm \frac{\sin(\beta t)}{\beta} = \sin(t) \pm \sin(\beta t) \pm \left(\frac{1}{\beta} - 1\right)\sin(\beta t) \tag{4.48}$$

for any β, we can uniformly approximate θ and ϕ in (4.47) by

$$\begin{aligned} \theta &= \frac{v}{2}\{\sin(t) - \sin(\sqrt{(1+2\alpha)}t)\} \\ \phi &= \frac{v}{2}\{\sin(t) + \sin(\sqrt{(1+2\alpha)}t)\}, \end{aligned} \tag{4.49}$$

incurring, at worst, an error of magnitude

$$1 - \frac{1}{\sqrt{(1+2\alpha)}} = \alpha + 0(\alpha^2). \tag{4.50}$$

Finally, (4.49) can be approximated by

$$\theta \doteq -v\cos(t)\sin(\alpha t/2) \tag{4.51}$$

and

$$\phi \doteq v\sin(t)\cos(\alpha t/2). \tag{4.52}$$

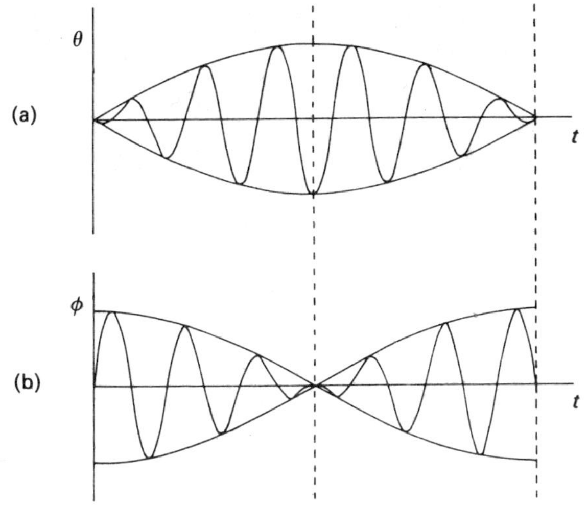

Fig. 4.11. Time dependence of angular displacements θ and ϕ given in (4.51) and (4.52) for coupled pendula. The envelopes (a) $\sin(\alpha t/2)$ for θ and (b) $\cos(\alpha t/2)$ for ϕ are out of phase by $\pi/2$ so that the amplitude of the oscillations in θ is a maximum when ϕ is almost zero.

The time dependence of θ and ϕ is shown in Fig. 4.11. For t near zero, pendulum B oscillates strongly while pendulum A shows only small amplitude oscillations. As t increases, the amplitude of the oscillations of B decays, while A oscillates with increasing vigour. At $t = \pi/\alpha$, pendulum B is stationary and A is oscillating with its maximum amplitude. The roles of A and B are now the reverse of that at $t = 0$ and the oscillations of A decline while those of B grow as t increases to $t = 2\pi/\alpha$. This phenomenon, which corresponds to a repeated exchange of energy between the pendula via the spring, is known as '*beats*'. It is a common feature of the motion of lightly coupled systems.

4.2 Affine models

In this section we consider models whose dynamical equations take the affine form,

$$\dot{\boldsymbol{x}} = \boldsymbol{A}\boldsymbol{x} + \boldsymbol{h}(t). \tag{4.53}$$

Such models can represent physical systems like those in Section 4.1 in the presence of time dependent external disturbances.

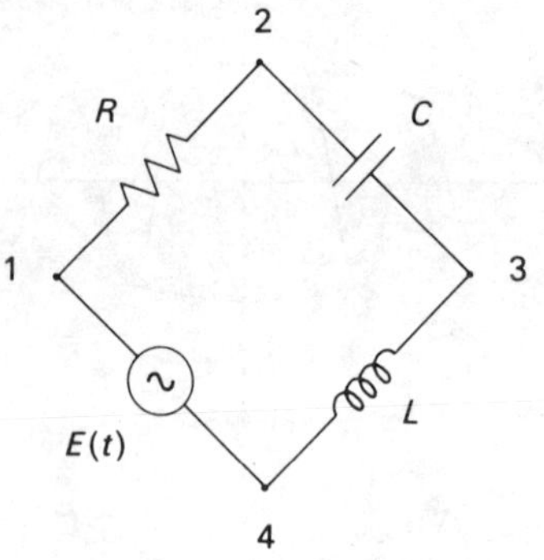

Fig. 4.12. Introduction of the generator as shown above replaces the linear dynamical equations (4.20) by an affine form.

The source of the disturbance or *forcing term*, $\boldsymbol{h}(t)$, depends on the physical system being modelled. For example, if the mass on the spring in Section 4.1.1 is subjected to an externally applied downward force $F(t)$ per unit mass, (4.3) becomes

$$\dot{x}_1 = x_2, \qquad \dot{x}_2 = -\omega_0^2 x_1 - 2kx_2 + F(t) \tag{4.54}$$

and

$$\boldsymbol{h}(t) = \begin{bmatrix} 0 \\ F(t) \end{bmatrix}. \tag{4.55}$$

Similarly, the inclusion of a generator in the LCR circuit of Section 4.1.2 (see Fig. 4.12) leads to

$$\boldsymbol{h}(t) = \begin{bmatrix} 0 \\ E(t)/LC \end{bmatrix}. \tag{4.56}$$

Let us consider the effect of a periodic forcing term

$$\boldsymbol{h}(t) = \begin{bmatrix} 0 \\ h_0 \cos \omega t \end{bmatrix} \tag{4.57}$$

on the damped harmonic oscillator discussed in Section 4.1.1 and 4.1.2.

4.2.1 *The forced harmonic oscillator*

Equation (2.92) gives

$$x(t) = \mathrm{e}^{At}\boldsymbol{x}_0 + \mathrm{e}^{At}\int_0^t \mathrm{e}^{-As}\boldsymbol{h}(s)\mathrm{d}s \tag{4.58}$$

for the solution of (4.53) which satisfies the initial condition $\boldsymbol{x}(0) = \boldsymbol{x}_0$. For $\boldsymbol{h}(t)$ given by (4.57), the integral in (4.58) can be evaluated by using the matrix equivalent of integration by parts (see Exercise 4.12). The result is

$$(\omega^2\boldsymbol{I} + \boldsymbol{A}^2)\int_0^t \mathrm{e}^{-As}\boldsymbol{h}(s)\mathrm{d}s = \boldsymbol{Ah}(0) - \mathrm{e}^{-At}\{\dot{\boldsymbol{h}}(t) + \boldsymbol{Ah}(t)\}. \tag{4.59}$$

Thus

$$\boldsymbol{x}(t) = \boldsymbol{x}_T(t) + \boldsymbol{x}_S(t), \tag{4.60}$$

where

$$\boldsymbol{x}_T(t) = \mathrm{e}^{At}\{\mathrm{x}_0 + (\omega^2\boldsymbol{I} + \boldsymbol{A}^2)^{-1}\boldsymbol{Ah}(0)\}$$
$$\boldsymbol{x}_S(t) = -(\omega^2\boldsymbol{I} + \boldsymbol{A}^2)^{-1}\{\dot{\boldsymbol{h}}(t) + \boldsymbol{Ah}(t)\}.$$

The first term $\boldsymbol{x}_T(t)$ in (4.60) is a particular solution of the *free* (undisturbed) harmonic oscillator;

$$\dot{\boldsymbol{x}} = \boldsymbol{Ax} \quad \text{with} \quad \boldsymbol{A} = \begin{bmatrix} 0 & 1 \\ -\omega_0^2 & -2k \end{bmatrix}. \tag{4.61}$$

In Section 4.1.1, we observed that the fixed point at the origin of (4.61) is asymptotically stable for all $k > 0$, so $\boldsymbol{x}_T(t) \to \boldsymbol{0}$ as $t \to \infty$. We say that $\boldsymbol{x}_T(t)$ is the *transient* part of the solution. The second term, $\boldsymbol{x}_S(t)$, in (4.60) is persistent and is said to represent the *steady state solution*; this remains after the 'transients' have died away.

4.2.2 *Resonance*

This phenomenon occurs when the frequency, ω, of the forcing term is near a special frequency determined by ω_0 and k. To see how this comes about we must examine $\boldsymbol{x}_S(t)$ more closely.

It is straightforward to show that

$$\omega^2\boldsymbol{I} + \boldsymbol{A}^2 = \begin{bmatrix} \omega^2 - \omega_0^2 & -2k \\ 2k\omega_0^2 & \omega^2 - \omega_0^2 + 4k^2 \end{bmatrix}, \tag{4.62}$$

and observe that

$$\begin{aligned} \det(\omega^2 \boldsymbol{I} + \boldsymbol{A}^2) &= (\omega^2 - \omega_0^2)^2 + 4k^2\omega^2 \\ &\neq 0 \quad \text{for } k > 0; \end{aligned} \tag{4.63}$$

so an inverse exists. This is given by

$$(\omega^2 \boldsymbol{I} + \boldsymbol{A}^2)^{-1} = \frac{1}{(\omega^2 - \omega_0^2)^2 + 4k^2\omega^2} \begin{bmatrix} \omega^2 - \omega_0^2 + 4k^2 & +2k \\ -2k\omega_0^2 & \omega^2 - \omega_0^2 \end{bmatrix}. \tag{4.64}$$

Finally we can conclude that

$$\begin{aligned} \boldsymbol{x}_S(t) &= \begin{bmatrix} x_{S1}(t) \\ x_{S2}(t) \end{bmatrix} \\ &= \frac{h_0}{(\omega^2 - \omega_0^2)^2 + 4k^2\omega^2} \begin{bmatrix} (\omega_0^2 - \omega^2)\cos\omega t + 2k\omega \sin\omega t \\ 2k\omega^2 \cos\omega t - (\omega_0^2 - \omega^2)\omega \sin\omega t \end{bmatrix}. \end{aligned} \tag{4.65}$$

Observe $x_{S2}(t) = \dot{x}_{S1}(t)$, as indeed it must since $\boldsymbol{x}(t) \to \boldsymbol{x}_S(t)$ as $t \to \infty$; so let us examine $x_{S1}(t)$ alone. This is simply a linear combination of $\cos\omega t$ and $\sin\omega t$ and it can therefore be written as

$$x_{S1}(t) = G(\omega)\cos(\omega t - \eta(\omega)) \tag{4.66}$$

where

$$G(\omega) = \frac{h_0}{[(\omega^2 - \omega_0^2)^2 + 4k^2\omega^2]^{1/2}} \tag{4.67}$$

and

$$\eta(\omega) = \tan^{-1}\left[\frac{2k\omega}{\omega_0^2 - \omega^2}\right]. \tag{4.68}$$

For a given physical system (i.e. fixed ω_0 and k), $G(\omega)$ has a maximum at $\omega = \omega_R$ given by

$$\omega_R = (\omega_0^2 - 2k^2)^{1/2} \tag{4.69}$$

provided $\omega_0^2 > 2k^2$. What is more,

$$G(\omega_R) = h_0/2k(\omega_0^2 - k^2)^{1/2} \tag{4.70}$$

and $G(\omega_R)/h_0 \to \infty$ as $k \to 0$. This means that if the physical system is sufficiently lightly damped, then the amplitude, $G(\omega_R)$, of the response is large, compared with that of the forcing term, h_0 (see Fig. 4.13). The

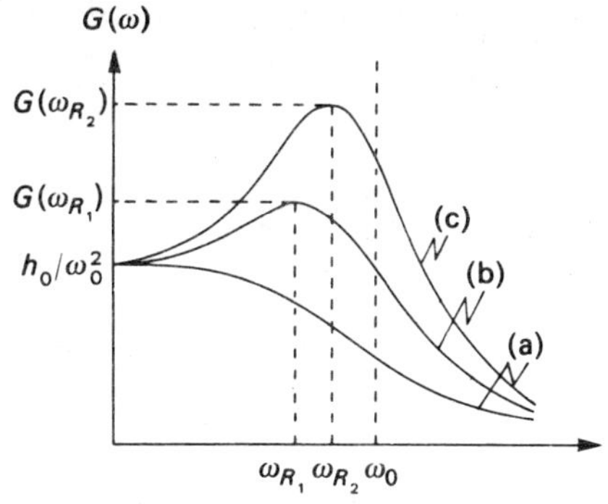

Fig. 4.13. Plots of $G(\omega)$ versus ω for: (a) $k = k_0$ where $\omega_0^2 < 2k_0^2$; (b) $k = k_1$ where $\omega_0^2 > 2k_1^2$: maximum at $\omega_{R_1} < \omega_0$; (c) $k = k_2 < k_1$ with maximum at ω_{R_2} where $\omega_{R_1} < \omega_{R_2} < \omega_0$. Observe the sharpening and shift of the maxima as k decreases.

system is said to *resonate* with the disturbance and ω_R is called the *resonant frequency*.

Resonances do have very important applications particularly in electronics where tuning devices are in common use. We can illustrate this idea by considering the series *LCR* circuit shown in Fig. 4.12 with $E(t) = E_0 \cos \omega t$. We have already observed in (4.20) and (4.56) that for this case

$$\omega_0^2 = 1/LC, \qquad 2k = R/L \quad \text{and} \quad \boldsymbol{h}(t) = \begin{bmatrix} 0 \\ (E_0/LC)\cos \omega t \end{bmatrix}. \tag{4.71}$$

The current flowing through the circuit in the steady state is $Cx_{S2}(t)$, i.e.

$$j(t) = -C\omega G(\omega) \sin(\omega t - \eta(\omega)) = j_0 \cos(\omega t + [\pi/2 - \eta(\omega)]). \tag{4.72}$$

This has amplitude

$$\begin{aligned} j_0 = C\omega G(\omega) &= \frac{E_0 \omega}{L\{[\omega^2 - (1/LC)]^2 + (R^2/L^2)\omega^2\}^{1/2}} \\ &= \frac{E_0}{\{R^2 + [\omega L - (1/\omega C)]^2\}^{1/2}}. \end{aligned} \tag{4.73}$$

The ratio $Z = E_0/j_0$ is called the *impedance* of the circuit; so for the series LCR circuit

$$Z = \{R^2 + [\omega L - (1/\omega C)]^2\}^{1/2} \tag{4.74}$$

In this case the resonant frequency is given by

$$\omega_R L - \frac{1}{\omega_R C} = 0 \quad \text{or} \quad \omega_R^2 = 1/LC. \tag{4.75}$$

At this frequency the impedance of the circuit is a minimum ($= R$) and for frequencies in this neighbourhood the impedance is small if R is small. For other frequencies the impedance increases with their separation from ω_R and the amplitudes of the associated currents would be correspondingly smaller. The circuit is said to be *selective*, favouring response in the neighbourhood of ω_R.

If, for example, the capacitance C can be varied then the resonant frequency can be changed. It is then possible to *tune* the circuit to provide strong current response at a desired frequency.

4.3 Non-linear models

In spite of the simplicity and undoubted success of linear models, they do have limitations. For example, the qualitative behaviour of a linear model in the plane must be one of a finite number of types. Linear models can have only one isolated fixed point; they can only produce persistent oscillations of a harmonic (or trigonometric) kind; such oscillations always come from centres in their phase portraits, etc. If the qualitative behaviour of the physical system to be modelled is not consistent with these limitations then a linear model will be inadequate.

In this section, we consider some problems from population dynamics and electrical circuit theory for which non-linear models are required.

4.3.1 *Competing species*

Two similar species of animal compete with each other in an environment where their common food supply is limited. There are several possible outcomes to this competition:

(a) species 1 survives and species 2 becomes extinct;
(b) species 2 survives and species 1 becomes extinct;
(c) the two species coexist;
(d) both species become extinct.

Each of these outcomes can be represented by an equilibrium state of the populations x_1 and x_2 of the two species. The differential equations used to model the dynamics of x_1 and x_2 are therefore required to have four isolated fixed points.

Consider the following non-linear dynamical equations:

$$\dot{x}_1 = (a - bx_1 - \sigma x_2)x_1, \qquad \dot{x}_2 = (c - vx_1 - dx_2)x_2, \tag{4.76}$$

where $a, b, c, d, \sigma, v > 0$. Observe that the per capita growth rate $\dot{x}_1/x_1 = (a - bx_1 - \sigma x_2)$ consists of three terms: the growth rate, a, of the isolated population; the 'intra'-species competition, $-bx_1$ (cf. Section 1.2.2); and the 'inter'-species competition, $-\sigma x_2$. A similar interpretation can be given for the terms c, $-dx_2$ and $-vx_1$ in $\dot{x}_2/x_2$.

A necessary condition for coexistence of the two species is that (4.76) has a fixed point with both populations greater than zero. Such a fixed point can only arise in (4.76) if the linear equations

$$bx_1 + \sigma x_2 = a, \qquad vx_1 + dx_2 = c \tag{4.77}$$

have a solution. We will assume that (4.77) has a unique solution in the positive quadrant of the x_1x_2-plane. The fixed point is then given by

$$\left(\frac{ad - \sigma c}{bd - v\sigma}, \frac{bc - av}{bd - v\sigma}\right), \tag{4.78}$$

where either:

$$bd < v\sigma \quad \text{with} \quad ad < \sigma c \quad \text{and} \quad bc < av; \tag{4.79}$$

or

$$bd > v\sigma \quad \text{with} \quad ad > \sigma c \quad \text{and} \quad bc > av. \tag{4.80}$$

The geometrical significance of these inequalities is illustrated in Fig. 4.14.

We will now consider the dynamics of x_1 and x_2 when (4.79) is satisfied; some other possibilities are dealt with in Exercise 4.16. We begin by examining the linearizations at the four fixed points. Let each

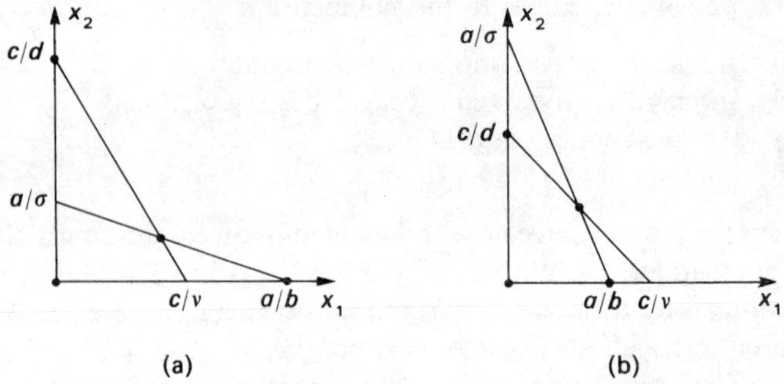

Fig. 4.14. Geometrical significance of (a) (4.79) and (b) (4.80). The fixed points of (4.76) are denoted by •

linearized system be denoted by $\dot{\boldsymbol{y}} = \boldsymbol{W}\boldsymbol{y}$, where $\boldsymbol{y}$ are local coordinates at the fixed point, then each fixed point and its corresponding $\boldsymbol{W}$ are given by:

$$(0,0), \qquad \begin{bmatrix} a & 0 \\ 0 & c \end{bmatrix}; \tag{4.81}$$

$$(0, c/d), \qquad \begin{bmatrix} a - \sigma c/d & 0 \\ -vc/d & -c \end{bmatrix}; \tag{4.82}$$

$$(a/b, 0), \qquad \begin{bmatrix} -a & -\sigma a/b \\ 0 & c - va/b \end{bmatrix}; \tag{4.83}$$

$$\left(\frac{\sigma c - ad}{v\sigma - bd}, \frac{av - bc}{v\sigma - bd}\right), \qquad \frac{1}{bd - v\sigma}\begin{bmatrix} b(\sigma c - ad) & \sigma(\sigma c - ad) \\ v(av - bc) & d(av - bc) \end{bmatrix} \tag{4.84}$$

The fixed points are all simple and their nature is determined by the eigenvalues of $\boldsymbol{W}$. For (4.81) these are $a > 0$ and $c > 0$ and, by the linearization theorem, the origin is an unstable node. The eigenvalues of $\boldsymbol{W}$ in (4.82) are also given by inspection, since $\boldsymbol{W}$ is triangular, but both are negative (cf. (4.79)). The same is true of (4.83) and in both cases the linearization theorem allows us to conclude that these fixed points are stable nodes or improper nodes (note that $a - \sigma c/d = -c$ or $c - va/b = -a$ is possible). Finally, the fixed point (4.84) is a saddle point, because

$$\det(\boldsymbol{W}) = \frac{(\sigma c - ad)(av - bc)}{bd - v\sigma} < 0, \tag{4.85}$$

by (4.79), and the eigenvalues of $\boldsymbol{W}$ must have opposite sign (cf. Fig. 2.7).

It is now apparent that coexistence is extremely unlikely in this model since only the two stable separatrices approach the saddle point as $t \to \infty$. The fixed points at $(0, c/d)$ and $(a/b, 0)$ are stable and correspond, respectively, to either species 1 or species 2 becoming extinct. The origin is an unstable node so there is no possibility of both populations vanishing since all trajectories move away from this point. Assuming that all initial states in the positive quadrant are equally likely, then by far the most likely outcome of the competition is that one or other of the species will die out. This result agrees with the 'Law of Competitive Exclusion'; namely, that when situations of this competitive kind occur in nature one or other of the species involved becomes extinct.

Further details of the evolution of the two species can be obtained by sketching the phase portrait for the model. Observe that the straight lines given in (4.77) are respectively the $\dot{x}_1 = 0$ and $\dot{x}_2 = 0$ isoclines. The sense of the vector field $((a - bx_1 - \sigma x_2)x_1,\ (c - vx_1 - dx_2)x_2)$ can be obtained by recognizing that $\dot{x}_1$ and $\dot{x}_2$ take positive and negative values as shown in Fig. 4.15. This information, along

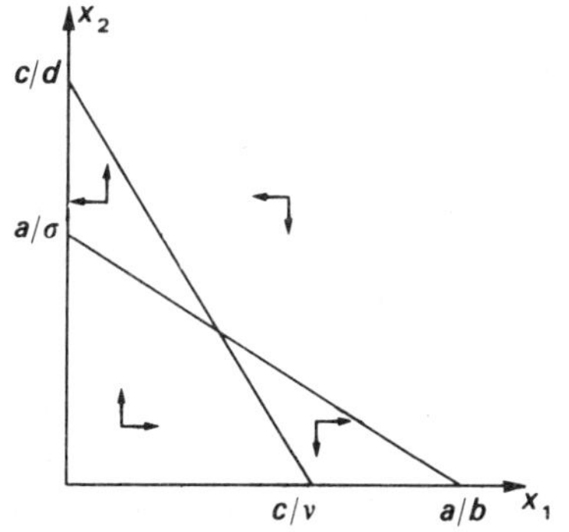

Fig. 4.15. The signs of $\dot{x}_1$ and $\dot{x}_2$ given by (4.76) in the positive quadrant of the x_1x_2-plane. The notation is as follows: ↳ means $\dot{x}_1, \dot{x}_2 > 0$; ↱ is $\dot{x}_1 > 0$, $\dot{x}_2 < 0$; ⬏ is $\dot{x}_1 < 0$, $\dot{x}_2 > 0$ and ↴ is $\dot{x}_1, \dot{x}_2 < 0$.

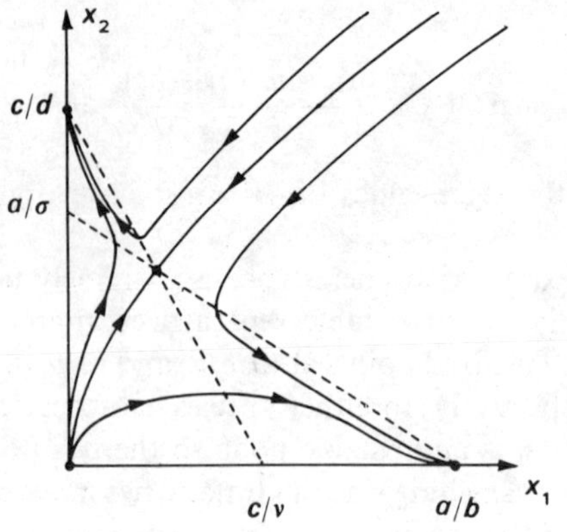

Fig. 4.16. Phase portrait for the competing species model (4.76) with $bd < \nu\sigma$, $ad < \sigma c$, $bc < a\nu$.

with the nature of the fixed points, is sufficient to construct a sketch of the phase portrait shown in Fig. 4.16.

Details such as whether trajectories leave the origin tangential to the x_1- or x_2-axis depend upon actual values of the eigenvalues. For instance, in Fig. 4.16 we have assumed $c < a$ so that trajectories are drawn tangential to the x_2-axis. Further such detail is dealt with in Exercise 4.15.

4.3.2 *Volterra–Lotka equations*

The most general oscillatory form that can arise from linear dynamical equations in the plane is

$$x_1(t) = R_1 \cos(\beta t + \theta_1), \qquad x_2(t) = R_2 \cos(\beta t + \theta_2), \tag{4.86}$$

corresponding to elliptical closed orbits in the phase portrait. These harmonic oscillations in (4.86) are very symmetric and, while R_i and θ_i ($i = 1, 2$) allow variation of amplitude and shift of origin, their shape remains unchanged.

In population dynamics, there are many examples of populations exhibiting oscillatory behaviour. However, these oscillations are markedly different from harmonic forms and non-linear dynamical equations are required to model them. The Volterra–Lotka equations are a particularly well-known example.

Consider a two species model in which one species is a predator that preys upon the other. Let x_1 and x_2 be the populations of prey and predator, respectively, and assume that there is no competition between individuals in either species. The per capita growth rate $\dot{x}_1/x_1$ of prey is taken as $a - bx_2$, $a, b > 0$; a is the growth rate in the absence of predators and $-bx_2$ allows for losses due to their presence. The predator population would decline in the absence of their food supply (i.e. the prey), so when $x_1 = 0$, $\dot{x}_2/x_2 = -c$, $c > 0$. However, successful hunting offsets this decline and we take $\dot{x}_2/x_2 = -c + dx_1$, $d > 0$, when $x_1 > 0$. Thus

$$\dot{x}_1 = (a - bx_2)x_1, \qquad \dot{x}_2 = (-c + dx_1)x_2, \tag{4.87}$$

where a, b, c, $d > 0$. These are the Volterra–Lotka equations.

The system (4.87) has two fixed points (0, 0) and $(c/d, a/b)$. The former is a saddle point with separatrices which coincide with the x_1- and x_2-axes; the latter being stable. The linearization at the non-trivial fixed point, $(c/d, a/b)$, is a centre and therefore the linearization theorem is unable to determine its nature.

The system (4.87) has a first integral

$$f(x_1, x_2) = x_1^c e^{-dx_1} x_2^a e^{-bx_2} = g(x_1)h(x_2) \tag{4.88}$$

(cf. Example 3.7.3). The functions $g(x_1)$ and $h(x_2)$ have the same form; each is positive for all arguments in $(0, \infty)$ and has a single maximum in this interval. These functions are plotted for typical values of their parameters in Fig. 4.17.

The maxima occur at $x_1 = c/d$ and $x_2 = a/b$, respectively, and thus $f(x_1, x_2)$ has a maximum of $g(c/d)h(a/b)$ at $(x_1, x_2) = (c/d, a/b)$. The

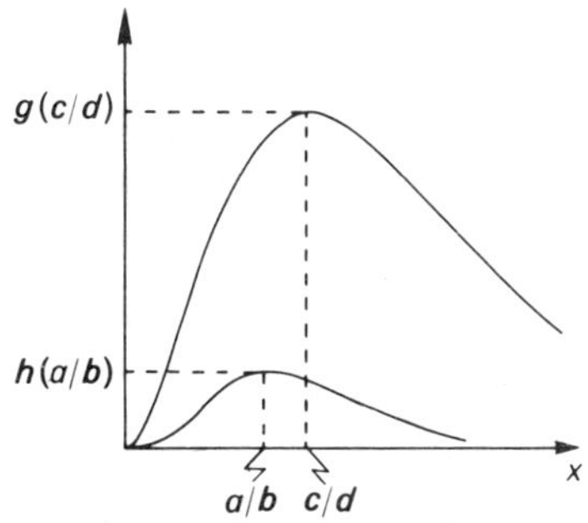

Fig. 4.17. Graphs of $g(x)$ and $h(x)$ for $a = 4.0$, $b = 2.5$, $c = 2.0$ and $d = 1.0$. Maxima occur at $x = c/d$ and $x = a/b$, respectively.

level curves of $f(x_1, x_2)$ are therefore closed paths surrounding $(c/d, a/b)$. The trajectories of (4.87) coincide with these curves and clearly $(c/d, a/b)$ is a centre.

The functions $g(x_1)$ and $h(x_2)$ are not symmetric about their extreme points and consequently the closed trajectories are certainly not ellipses. The sharp rise followed by slower fall shown by both g and h (see Fig. 4.17) means that the trajectories take the form shown in Fig. 4.18.

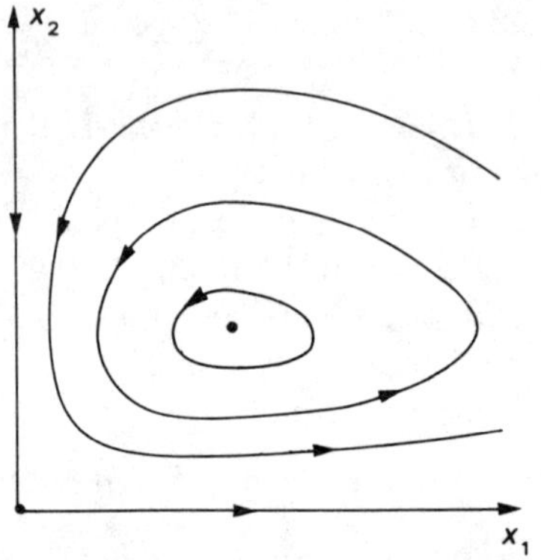

Fig. 4.18. Phase portrait for the Volterra–Lotka equations $\dot{x}_1 = (a - bx_2)x_1$, $\dot{x}_2 = (-c + dx_1)x_2$ with $a = 4.0$, $b = 2.5$, $c = 2.0$, $d = 1.0$.

The non-elliptic form of the trajectories of the non-linear centre is reflected in the non-harmonic nature of the oscillations in the populations (see Fig. 4.19).

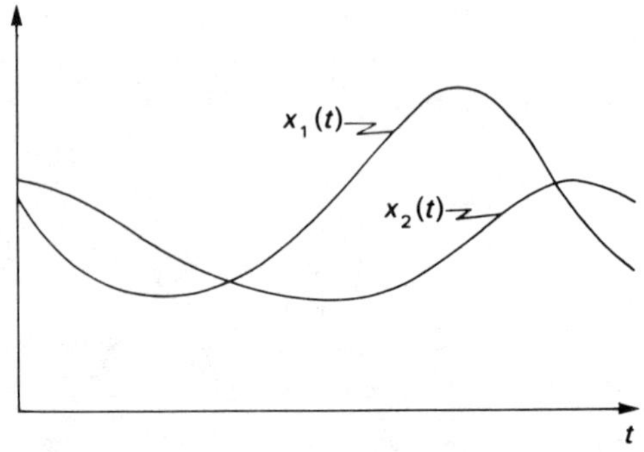

Fig. 4.19. Oscillations in x_1 and x_2 obtained from the phase portrait in Fig. 4.18.

4.3.3 *The Holling–Tanner model*

Although persistent non-harmonic oscillations in dynamical variables can be modelled by systems with centres in their phase portraits, the stability of such oscillations to perturbations of the model is suspect. We saw in Section 4.1 that the centre for the harmonic oscillator was destroyed by even the weakest of damping forces; becoming a focus. The same is true of the simple pendulum (see Exercise 4.25). Similarly, suppose the Volterra–Lotka equations (4.87) are modified to include the additional effects of competition within the species. The resulting system no longer has a centre in its phase portrait (see Exercise 4.17) and the population oscillations decay. This tendency to be very easily destroyed is inherent in the make-up of centres. They are said to lack *structural stability*.

An alternative way for persistent oscillations to occur in non-linear systems is by the presence of a stable *limit cycle* in their phase portraits (cf. Section 3.8). The limit cycle *is* structurally stable and is consequently a 'more permanent' feature of a phase portrait; it is not likely to disappear as a result of relatively small perturbations of the model. Models that are insensitive to perturbations are said to be *robust*. Since most models are idealizations which focus attention on certain central variables and interactions, this kind of stability is very important.

The two-species prey–predator problem is obviously an idealization of the kind mentioned above and the Volterra–Lotka model is not robust. It is therefore questionable whether (4.87) contains the true mechanism for population oscillations.

A model which provides structurally stable population oscillations for the prey–predator problem is the Holling–Tanner model. The dynamical equations are

$$\begin{aligned} \dot{x}_1 &= r(1 - x_1 K^{-1})x_1 - w x_1 x_2 (D + x_1)^{-1} \\ \dot{x}_2 &= s(1 - J x_2 x_1^{-1})x_2 \end{aligned} \tag{4.89}$$

with $r, s, K, D, J > 0$.

The rate of growth of prey $\dot{x}_1$ is the difference of two terms:

(a) $r(1 - x_1 K^{-1})x_1$, which gives the growth of the prey population in the absence of predators. This includes competition between prey for some limiting resource (see Section 1.2.2);

(b) $w x_1 x_2 (D + x_1)^{-1}$, which describes the effect of the predators.

To explain the form of (b) it is convenient to think of the effect of predators on $\dot{x}_1$ in terms of a *predation rate*. This is the number of prey killed per predator per unit time. In (4.87) the predation rate is bx_1. This means that the number of prey killed per predator per unit time increases indefinitely with prey population.

A more reasonable assumption is that there is an upper limit to the predation rate, e.g. when the predator's appetite is satisfied. This is taken into account in (b) where the predation rate is $wx_1(D+x_1)^{-1}$ (see Fig. 4.20).

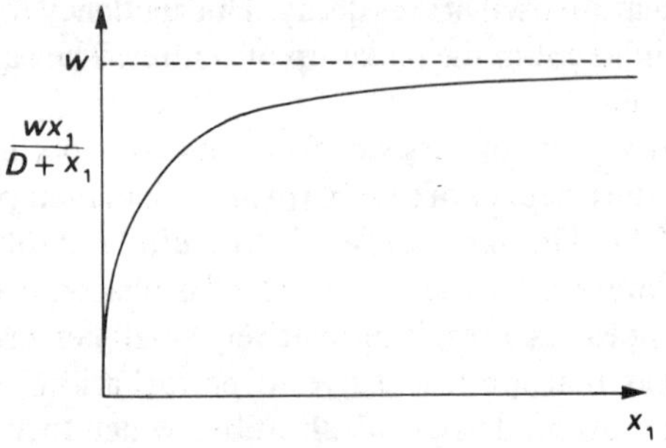

Fig. 4.20. The predation rate for the Holling–Tanner model. The slope at $x_1 = 0$ is w/D and the saturation value is w.

The rate of growth of the predator population, $\dot{x}_2$, is obtained by viewing the prey as a scarce resource. We will assume that the number of prey required to support one predator is J. Thus if the prey population is x_1 it can support no more than x_1/J predators. We can ensure that the predator population does not exceed its environmental limits by taking (cf. (1.18))

$$\dot{x}_2 = x_2\left(s - \frac{sJ}{x_1}x_2\right).$$

This is the equation given in (4.89).

The $\dot{x}_1 = 0$ and $\dot{x}_2 = 0$ isoclines of (4.89) are shown in Fig. 4.21. As can be seen, there is a fixed point with x_1 and x_2 both greater than zero; at $(x_1, x_2) = (x_1^*, x_2^*)$, say. To determine the nature of this fixed point it is convenient to scale the variables in (4.89) by x_1^*. We take

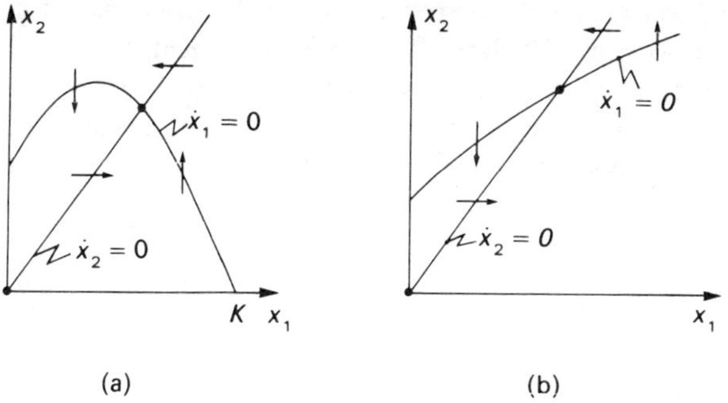

Fig. 4.21. Isoclines for (4.89) are given by: $\dot{x}_1 = 0$ on $x_2 = rw^{-1}(1 - x_1 K^{-1})(D + x_1)$, which has a maximum at $x_1 = (K - D)/2$; $\dot{x}_2 = 0$ on $x_2 = J^{-1}x_1$. Intersection of these isoclines can take place with (a) $x_1 > (K - D)/2$ or (b) $x_1 < (K - D)/2$.

$y_1 = x_1/x_1^*$ and $y_2 = x_2/x_1^*$ and obtain

$$\begin{aligned} \dot{y}_1 &= r(1 - y_1 k^{-1})y_1 - w y_1 y_2 (d + y_1)^{-1} \\ \dot{y}_2 &= s(1 - J y_2 y_1^{-1}) y_2 \end{aligned} \tag{4.90}$$

where $k = K/x_1^*$ and $d = D/x_1^*$. It follows that the fixed point in the $y_1 y_2$-plane has coordinates

$$\begin{aligned} (y_1^*, y_2^*) &= (1, rw^{-1}(1 - k^{-1})(1 + d)) \\ &= (1, J^{-1}). \end{aligned} \tag{4.91}$$

The coefficient matrix of the linearized system at (y_1^*, y_2^*) is then

$$W = \begin{bmatrix} r(-k^{-1} + w(rJ)^{-1}(1 + d)^{-2}) & -w(1 + d)^{-1} \\ sJ^{-1} & -s \end{bmatrix}. \tag{4.92}$$

Observe that

$$\det(W) = rs(k^{-1} + wd(rJ)^{-1}(1 + d)^{-2}) > 0 \tag{4.93}$$

so that (y_1^*, y_2^*) is never a saddle point. However, substituting for $w(rJ)^{-1}$ from (4.91),

$$\begin{aligned} \mathrm{Tr}(W) &= r(w(rJ)^{-1}(1 + d)^{-2} - k^{-1}) - s \\ &= r(k - d - 2)(k(1 + d))^{-1} - s, \end{aligned} \tag{4.94}$$

which can be either positive or negative.

Let ϕ_t be the evolution operator for (4.90) and consider $\phi_t(k, 0_+)$ for $t > 0$. As Fig. 4.22 shows, this trajectory must move around the fixed point and intersect the $\dot{x}_1 = 0$ isocline at A. The set S enclosed by this trajectory and the part of the $\dot{x}_1 = 0$ isocline between A and $(k, 0)$ is positively invariant.

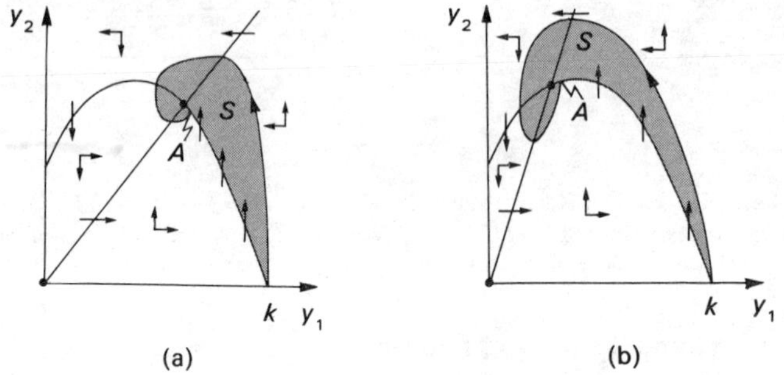

Fig. 4.22. Positively invariant sets S containing the fixed point (y_1^*, y_2^*), bounded by the trajectory $\{\phi_t(k, 0_+) \mid t > 0\}$ and part of the $\dot{x}_1 = 0$ isocline: (a) $x_1^* > (K - D)/2$; (b) $x_1^* < (K - D)/2$.

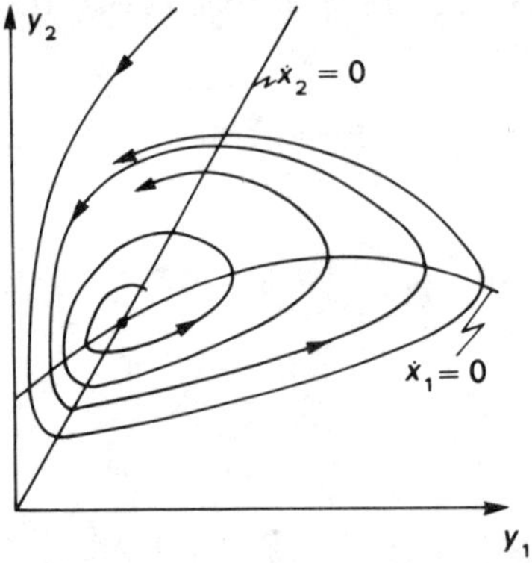

Fig. 4.23. Phase portrait for (4.90) with $r = 1.0, s = 0.2, k = 7.0, d = 1.0$; showing the stable limit cycle of the Holling–Tanner model.

If $\mathrm{Tr}(\boldsymbol{W}) > 0$, so that (y_1^*, y_2^*) is an unstable fixed point then the linearization theorem ensures that there is a neighbourhood N of (y_1^*, y_2^*) in S such that $S \setminus N$ is positively invariant. However, there are no fixed points in $S \setminus N$ and by the corollary to Theorem 3.9.1 this set must contain a closed orbit. Thus, for those sets of parameters (r, s, k, d) for which

$$s < \frac{r(k-d-2)}{k(1+d)}, \tag{4.95}$$

the phase portrait of (4.90) contains a stable limit cycle (see Fig. 4.23). In Exercise 4.19, it is shown that this can only occur if (y_1^*, y_2^*) lies to the left of the peak in the $\dot{x}_1 = 0$ isocline. Thus the phase portrait corresponding to Fig. 4.22(a) has no limit cycle; (y_1^*, y_2^*) is simply a stable focus.

4.4 Relaxation oscillations

4.4.1 *Van der Pol oscillator*

Consider the modification of the series *LCR* circuit shown in Fig. 4.24. Suppose that the 'black box' *B* is a circuit element (or

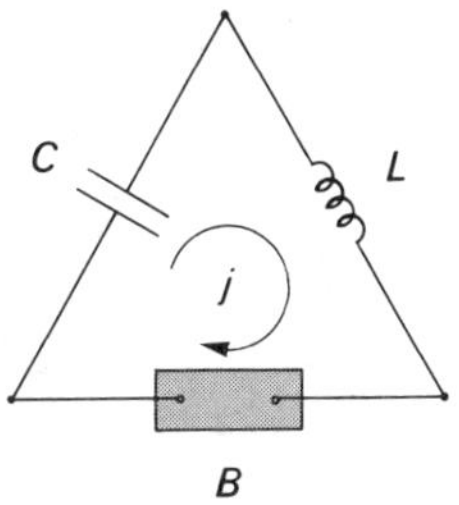

Fig. 4.24. A modification of the series *LCR* circuit.

collection of circuit elements) with a voltage–current relationship like that shown in Fig. 4.25(a). The black box is said to be a non-linear resistor with the cubic *characteristic*,

$$v_B = f(j_B) = j_B(\tfrac{1}{3}j_B^2 - 1). \tag{4.96}$$

Let v_L, v_B and $-v_C$ be the potential differences across the elements in Fig. 4.24 *in the direction of the current j*. The dynamical equations for the circuit are

$$L\frac{dj}{dt} = v_C - f(j); \qquad C\frac{dv_C}{dt} = -j, \tag{4.97}$$

where we have used Kirchhoff's voltage law to eliminate v_L. Now let

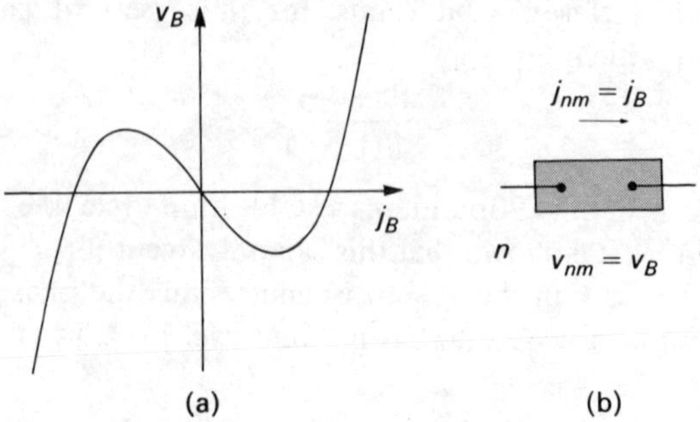

Fig. 4.25. The voltage–current 'characteristic' (a) of the black box in Fig. 4.24. With the notation in (b) $v_B = f(j_B) = j_B(\frac{1}{3}j_B^2 - 1)$.

$L^{-1}t \to t$, $x_1 = j$, $x_2 = v_C$, $L/C = \eta$, to obtain

$$\dot{x}_1 = x_2 - f(x_1), \qquad \dot{x}_2 = -\eta x_1. \tag{4.98}$$

The system (4.98) has a fixed point at $x_1 = x_2 = 0$ and the isoclines shown in Fig. 4.26.

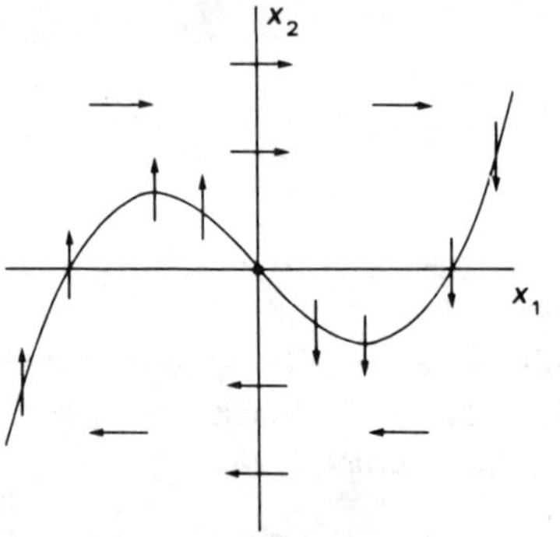

Fig. 4.26. Isoclines for (4.98): $\dot{x}_1 = 0$ on $x_2 = f(x_1)$; $\dot{x}_2 = 0$ on x_2-axis. The sense of the vector field is shown for small η.

Suppose now that η is so small (L small compared with C) that $\dot{x}_2$ can be neglected, relative to $\dot{x}_1$, at all points of the phase plane that are not in the immediate neighbourhood of the $\dot{x}_1 = 0$ isocline. This means that the vector field $(x_2 - f(x_1), -\eta x_1)$ is directed essentially horizontally in Fig. 4.26 except near the curve

$$x_2 = f(x_1) = \frac{x_1^3}{3} - x_1. \tag{4.99}$$

It follows that the phase portrait must be as shown in Fig. 4.27. As can be seen, all initial values of (x_1, x_2) lead to trajectories which home in on the limit cycle *ABCD* at the first opportunity. At points away from the cubic (4.99) $\dot{x}_1$ is comparatively large so that the phase point moves rapidly to the neighbourhood of the characteristic and slowly follows it.

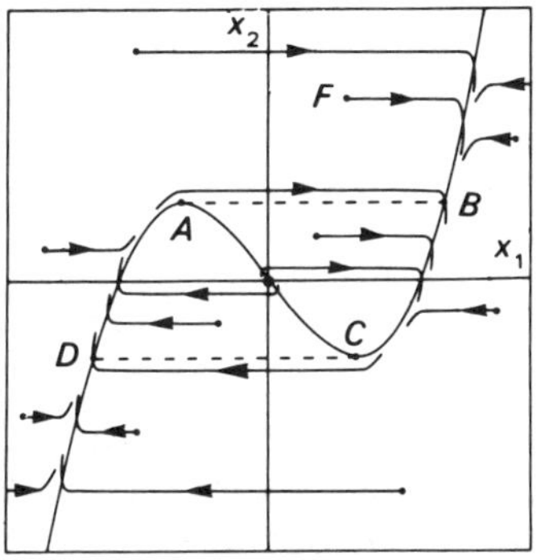

Fig. 4.27. Phase portrait for (4.98) when $\eta \to 0$. All trajectories rapidly home in on the limit cycle *ABCD*.

The fast movement on *AB*, *CD* and the slow movement on *BC*, *DA* leads to an 'almost discontinuous' wave form for the persistent oscillations of the current $x_1(t)$ (see Fig. 4.28). Oscillations of this kind are called *relaxation oscillations.*

The system (4.98) does not only have a limit cycle when $\eta \to 0$. For example, when $\eta = 1$ it is a special case of the *Van der Pol equation*

$$\ddot{x} + \varepsilon(x^2 - 1)\dot{x} + x = 0, \qquad \varepsilon > 0, \tag{4.100}$$

when $\varepsilon = 1$. This equation has the form

$$\ddot{x} + g(x)\dot{x} + h(x) = 0 \tag{4.101}$$

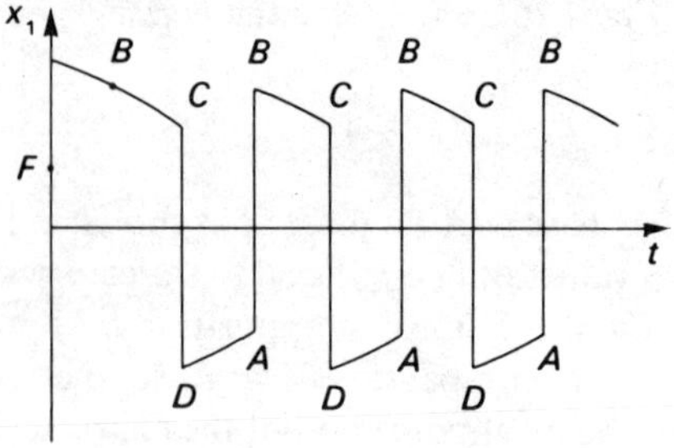

Fig. 4.28. Relaxation oscillations from the trajectory starting at F in Fig. 4.27. Sections marked CD and AB are almost vertical corresponding to rapid motion between these points in the phase plane.

which is equivalent to the system

$$\dot{x}_1 = x_2 - \int^{x_1} g(u)\mathrm{d}u; \qquad \dot{x}_2 = -h(x_1) \tag{4.102}$$

(cf. Exercise 1.27). Equations of the form (4.101) are sometimes called *Liénard equations* (see Section 5.1) and the coordinates in (4.102) are said to define the *Liénard plane*. The Liénard representation of the Van der Pol equation is

$$\dot{x}_1 = x_2 - \varepsilon\left(\frac{x_1^3}{3} - x_1\right), \qquad \dot{x}_2 = -x_1 \tag{4.103}$$

which is (4.98) when $\varepsilon = \eta = 1$.

The Van der Pol equation can be thought of as an extension of the harmonic oscillator equation (4.2). The difference is that it has a non-linear damping coefficient, $\varepsilon(x^2 - 1)$, instead of the constant $2k$ in (4.2).

The Van der Pol oscillator has a unique stable limit cycle for all values of ε. When $\varepsilon \to \infty$, this limit cycle takes the shape shown in Fig. 4.27 and relaxation oscillations result (see Exercise 4.28). When $\varepsilon \to 0$, the limit cycle is approximately circular with radius two (see Exercise 4.27). As ε increases from zero the limit cycle distorts smoothly from one of these extremes to the other (see Fig. 4.29).

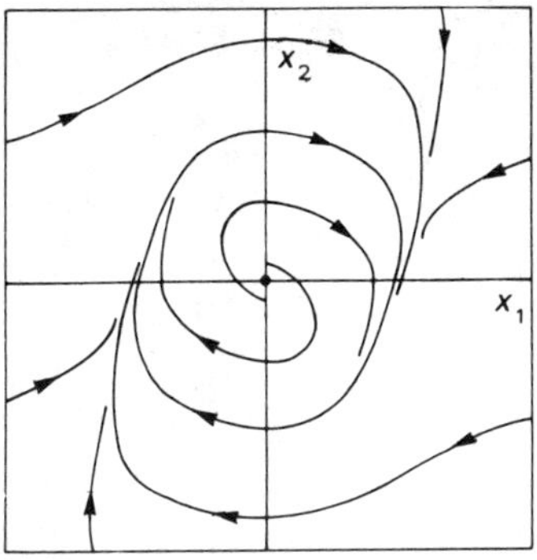

Fig. 4.29. Limit cycle of Van der Pol oscillator in the Lienard plane: $\varepsilon = 1.0$

4.4.2 *Jumps and regularization*

If we let $L = 0$ in the circuit equations (4.97) the differential equation

$$L\frac{dj}{dt} = v_C - f(j) \tag{4.104}$$

is replaced by the algebraic equation

$$v_C = f(j). \tag{4.105}$$

Thus the system of two dynamical equations (4.97) with $(j, v_C) \in \mathbb{R}^2$ is replaced by a single dynamical equation

$$C\frac{dv_C}{dt} = C\dot{v}_C = -j \tag{4.106}$$

defined on $\{(j, v_C) | v_C = f(j)\}$.

Geometrically, (4.106) means that the dynamics of the circuit are confined to the curve (4.105) in the j, v_C-plane. Equation (4.106) implies that

$$\dot{v}_C \lessgtr 0 \quad \text{for} \quad j \gtrless 0, \tag{4.107}$$

and so the states of the circuit must evolve along the cubic characteristic as shown in Fig. 4.30. For any starting point on $v_C = f(j)$ the state of the circuit evolves to one of the two points A or C. These points are *not* fixed points, clearly $\dot{v}_C = -j/C \neq 0$ at both;

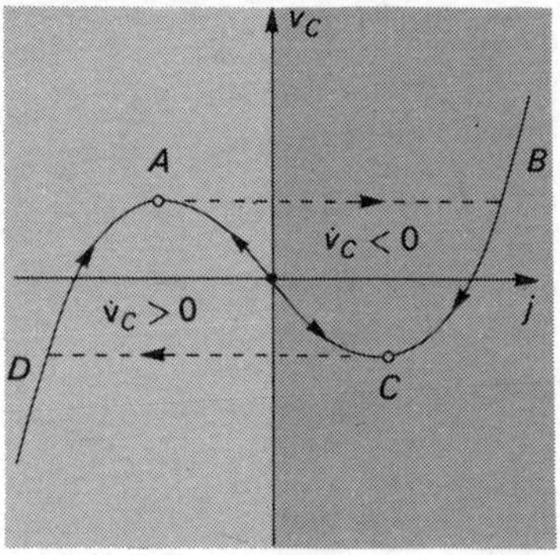

Fig. 4.30. Equation (4.107) means that v_C is increasing for $j < 0$ and decreasing for $j > 0$. Thus, for any initial point on $v_C = f(j)$, the state of the circuit evolves either to point A or to point C.

however they are special in that

$$\frac{\mathrm{d}j}{\mathrm{d}t} = -j\left(C\frac{\mathrm{d}f}{\mathrm{d}j}\right)^{-1} \tag{4.108}$$

is undefined (since $\mathrm{d}f/\mathrm{d}j$ vanishes).

Without the analysis of (4.97) given in Section 4.4.1, we would have difficulty in deciding what happens when A or C is reached. As it is, we can reasonably assume that j changes *discontinuously*; instantaneously 'jumping' to B or D, respectively, while v_C remains constant. Thus, relaxation oscillations occur in $j(t)$, but the changes are now genuinely discontinuous.

When a constrained system of dynamical equations is asymptotically related to an unconstrained system of higher dimension, the unconstrained system is called the *regularization* of the constrained one. Thus, (4.97) is the regularization of (4.106).

Sometimes constrained dynamical equations and assumptions of 'jumps' in the variables occur in the natural formulation of a model. In such cases, the regularization can represent a refinement of the model. We can illustrate this by the following example involving an electrical circuit containing a 'neon tube'.

A neon tube, or 'neon', is essentially a glass envelope containing an inert gas (neon) which is fitted with two electrodes. As the potential difference between these electrodes is increased from zero, no current flows until a threshold voltage v_1 is reached. At this voltage the tube suddenly conducts (the inert gas ionizes) and the current jumps to a non-zero value. Thereafter the current increases essentially linearly with increasing voltage. This is the portion $0 \to A \to B \to E$ of the voltage–current characteristic for the neon shown in Fig. 4.31.

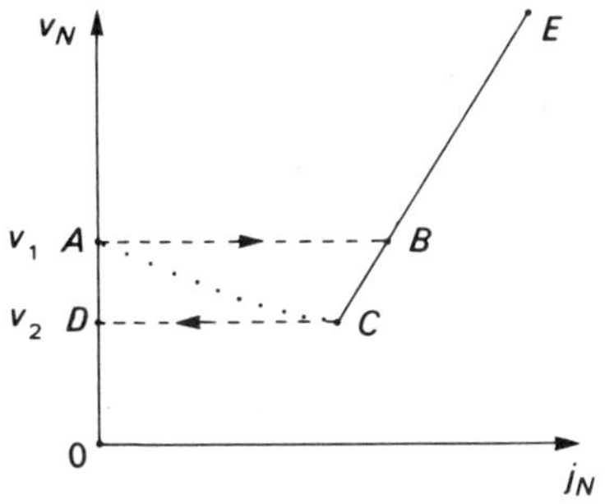

Fig. 4.31. An idealised voltage–current characteristic of a neon-tube. Observe the similarity between this and the cubic characteristic; both are folded curves.

When the potential difference is reduced from its value at E, however, the tube does *not* stop conducting when voltage v_1 is reached. Instead, the current continues to decrease with voltage until v_2 ($< v_1$) is reached; here the current falls sharply to zero. This is the path $E \to B \to C \to D$ on Fig. 4.31. This behaviour is completely reproducible if the voltage is again increased.

There is also physical evidence for a section of characteristic connecting C and A, shown dotted in Fig. 4.31. This does not appear in the simple steady voltage experiment described above and need not concern us greatly here. However, it does help to emphasize the 'folded' nature of the neon characteristic.

Example 4.4.1. Show that the dynamics of the circuit shown in Fig. 4.32 are governed by a constrained dynamical equation describing the time dependence of the voltage, v_C, across the capacitor.

Suppose that a small inductor, L, is introduced between nodes 2 and 3 in the figure, what effect does this have on the dynamical equations?

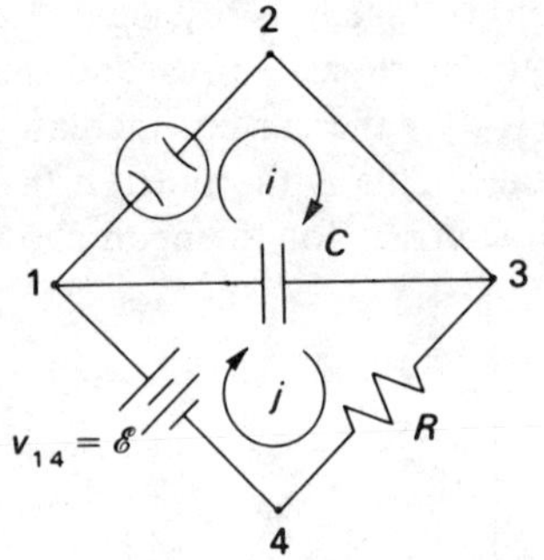

Fig. 4.32. A rudimentary time-base circuit (see Section 4.5.1).

Solution. Let $v_{12} = v_N, v_{13} = v_C$ and $v_{34} = v_R$ then, with the notation given in Fig. 4.32, we have

$$v_R + v_C = \mathscr{E} = jR + v_C \tag{4.109}$$

and

$$v_C = v_N = f(i), \tag{4.110}$$

where $f(i)$ is the voltage–current characteristic of the neon tube. Furthermore,

$$C\dot{v}_C = j - i. \tag{4.111}$$

Thus the behaviour of the circuit is determined by

$$C\dot{v}_C = \frac{\mathscr{E} - v_C}{R} - i \quad \text{with} \quad v_C = f(i), \tag{4.112}$$

which is a constrained dynamical equation.

If the inductor, L, is introduced between nodes 2 and 3, Equation (4.110) becomes

$$v_L = v_C - v_N, \tag{4.113}$$

where $v_L = v_{23}$. It then follows that the constraint $v_C = f(i)$ is replaced by the *new* dynamical equation

$$L\frac{\mathrm{d}i}{\mathrm{d}t} = v_C - f(i). \tag{4.114}$$

This equation, along with the differential equation in (4.112), forms an unconstrained system which is the regularization of (4.112). □

Jumps are built into the dynamics of the capacitor voltage in Example 4.4.1 by the characteristic of the neon. Suppose, v_C is initially zero; the battery, $\mathscr{E}$, charges the capacitor, C, through the resistor, R. The voltage across the neon tube, v_N, is equal to v_C. The current, i, through the neon is zero until $v_N = v_C = v_1$ ($0 \to D \to A$ on Fig. 4.31) when the neon conducts (i.e. there is a jump in i). Provided R is sufficiently large, the neon represents a low resistance path and the capacitor discharges through it. This means that v_C falls, however, the neon continues to conduct until $v_C = v_N = v_2$ when i jumps back to zero ($B \to C \to D$ on Fig. 4.31). The battery then charges the capacitor again. The jumps arise because v_C must be equal to $f(i)$.

The small inductance, L, changes all this. The dynamics are defined on the whole i, v_C-plane and the jumps are smoothed out as $ABCD$ in Fig. 4.31 becomes a limit cycle.

4.5 Piecewise modelling

In Section 4.4 we have considered models in which the dynamical equations are constrained by functional relationships between the dynamical variables. Another, apparently different type of constraint, arises in models which involve different dynamical equations in certain intervals of time or regions of the dynamical variables. Such models can often be thought of as being constructed in pieces and frequently give rise to piecewise continuous or piecewise differentiable time-dependences in the variables.

4.5.1 *The jump assumption and piecewise models*

At first sight these models might appear to be completely separate from the models with constraints considered in Section 4.4. However, this is not always the case, as the following example shows.

Example 4.5.1. Show that the constrained dynamical equation (4.112), i.e.

$$C\dot{v}_C = \frac{\mathscr{E} - v_C}{R} - i \quad \text{with} \quad v_C = f(i)$$

obtained in Example 4.4.1, is equivalent to a model involving two different dynamical equations in different intervals of time.

Refer to the neon tube characteristic shown in Fig. 4.31 and assume that on CE $v_N = rj_N$. Use the dynamical equations you have developed to find oscillatory solutions for $v_C(t)$ if $v_C = v_2$ and $i = 0$ when $t = 0$.

Solution. Consider the significance of the jump assumption associated with the dynamics of (4.112). If the portion CE of the characteristic shown in Fig. 4.31 is given by

$$j_N = \phi(v_N), \tag{4.115}$$

then the jump assumption implies that

$$i = \begin{cases} 0, & \text{on } OA \\ \phi(v_C), & \text{on } CE. \end{cases} \tag{4.116}$$

Thus, when the state of the circuit corresponds to a point on OA the dynamical equation is

$$C\dot{v}_C = \frac{\mathscr{E} - v_C}{R}. \tag{4.117}$$

However, when the state corresponds to a point on CE, the dynamical equation is

$$C\dot{v}_C = \frac{\mathscr{E} - v_C}{R} - \phi(v_C). \tag{4.118}$$

If $v_C(\mathrm{O}) = v_2$ and $i(\mathrm{O}) = 0$, the initial state of the system is at D on OA and the dynamics are given by (4.117). This is a linear (strictly affine) equation with solution

$$v_C(t) = \mathscr{E}[1 - \exp(-t/CR)] + v_2\exp(-t/CR). \tag{4.119}$$

The voltage $v_C = v_N$ therefore increases towards $\mathscr{E}$.

For oscillations to take place we require $\mathscr{E} > v_1$, so that after time T_1, say, the neon begins to conduct. The state of the circuit jumps to $B \in CE$ and the dynamical equation switches to (4.118). If we take $\phi(v_C) = v_C/r$, then (4.118) becomes

$$\dot{v}_C + \frac{v_C}{C\bar{R}} = \frac{\mathscr{E}}{CR}, \tag{4.120}$$

where $\bar{R}^{-1} = R^{-1} + r^{-1}$. This linear differential equation has solution

$$v_C = \frac{\mathscr{E}\bar{R}}{R}[1-\exp(-\tau/C\bar{R})] + v_1\exp(-\tau/C\bar{R}) \qquad (4.121)$$

where $\tau = t - T_1$.

For oscillations to occur $\mathscr{E}\bar{R}/R$ must be smaller than v_2 so that v_C now falls as t increases, finally reaching v_2 at $\tau = T_2$, say (see Fig. 4.33). At this voltage the neon stops conducting and the state jumps again to D where (4.117) takes over again and v_C rises once more. □

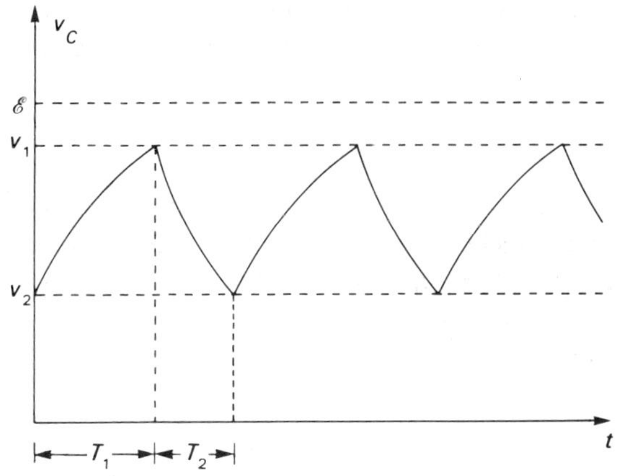

Fig. 4.33. Non-linear 'saw-tooth' oscillations obtained from the linear differential equations (4.117) and (4.118)

The voltage across the capacitor in Example 4.5.1 increases to v_2 at a rate determined by R (see (4.119)) and 'relaxes' back to v_1 in a way determined by $\bar{R}$ (see (4.121)). If $R \gg r$, $\bar{R} \simeq r$ and the relaxation takes place much more rapidly than the growth of v_C. If T_2 is sufficiently small compared with T_1 the return to v_2 is essentially instantaneous. Furthermore if $v_1 - v_2$ is small enough compared with $\mathscr{E}$, v_C is essentially linear for the rise from v_2 to v_1.

These are just the properties required for the time base of an oscilloscope. The x-plates are connected across the capacitor and the signal to be displayed is applied across the y-plates. The electron beam is therefore drawn steadily across the screen and returned rapidly to its starting point, once in every cycle. Synchronization of the period of v_C and the signal allows the latter to be displayed as a stationary wave form.

This example is particularly interesting because the non-harmonic, saw-tooth oscillations are clearly a non-linear phenomenon and yet they are modelled successfully by using the essentially *linear* differential equations (4.117) and (4.118) in the appropriate time intervals.

The two types of solution are joined continuously at $t = T_1, t = T_1 + T_2$, etc., although the slope $\dot{v}_C$ is discontinuous at these points. This discontinuity arises because of the jumps which take place in the phase plane shown in Fig. 4.31. The equations (4.117) and (4.118) are defined on the separated sets DA and BC of this figure and it is this separation which gives rise to the jumps and hence the discontinuity of $\dot{v}_C$.

4.5.2 *A limit cycle from linear equations*

There are many models in which non-linear phenomena are constructed by piecing together linear dynamical equations. Perhaps the most intriguing example of this kind is the construction of a limit cycle from two damped, harmonic oscillator equations (Andronov and Chaikin, 1949).

The system is defined by

$$\ddot{x} + 2k\dot{x} + \omega_0^2 x = \omega_0^2 g, \qquad \dot{x} > 0 \tag{4.122}$$

$$\ddot{x} + 2k\dot{x} + \omega_0^2 x = 0, \qquad \dot{x} \leqslant 0 \tag{4.123}$$

where g is a real constant. We have already examined the possible solutions to (4.123) in Section 4.1. For our purpose here it will be necessary to assume that $0 < k < \omega_0$ so that the system undergoes damped, free oscillations of the form

$$x(t) = R\mathrm{e}^{-kt} \cos(\beta t + \theta) \tag{4.124}$$

where $\beta = (\omega_0^2 - k^2)^{1/2}$. In the phase plane defined by $x_1 = x, x_2 = \dot{x}$ the phase portrait has a fixed point at (0, 0) which is a stable focus.

The solutions to (4.122) are easily obtained from the same information by recognizing that this equation can be written as

$$\ddot{y} + 2k\dot{y} + \omega_0^2 y = 0 \tag{4.125}$$

with $y = x - g$. It follows that (4.122) has solutions of the form

$$x(t) = g + \bar{R}\mathrm{e}^{-kt} \cos(\beta t + \bar{\theta}) \tag{4.126}$$

and its phase portrait in the x_1x_2-plane has a stable focus at $(g, 0)$.

In order to show that the system defined by (4.122) and (4.123) has a limit cycle in its phase portrait, we will first prove that a closed trajectory can be found. Let us suppose that at some time $t = t_1$ the phase point is at $(x_1(t_1), 0)$, with $x_1(t_1) > 0$; it moves according to (4.123) and spirals about the origin. At time $t_1 + \pi/\beta$ it again encounters the x_1-axis on the other side of the origin (i.e. it follows the segment of trajectory from P to Q in the lower half plane of Fig. 4.34). Equation (4.124) gives

$$x_1(t_1) = R\exp(-kt_1)\cos(\beta t_1 + \theta) \tag{4.127}$$

and

$$\begin{aligned} x_1\left(t_1 + \frac{\pi}{\beta}\right) &= R\exp\left[-k\left(t_1 + \frac{\pi}{\beta}\right)\right]\cos(\beta t_1 + \pi + \theta) \\ &= -R\exp(-kt_1)\exp\left(-\frac{k\pi}{\beta}\right)\cos(\beta t_1 + \theta). \end{aligned} \tag{4.128}$$

Thus

$$x_1\left(t_1 + \frac{\pi}{\beta}\right) = -\rho x_1(t_1) \tag{4.129}$$

with $\rho = \exp(-k\pi/\beta) < 1$.

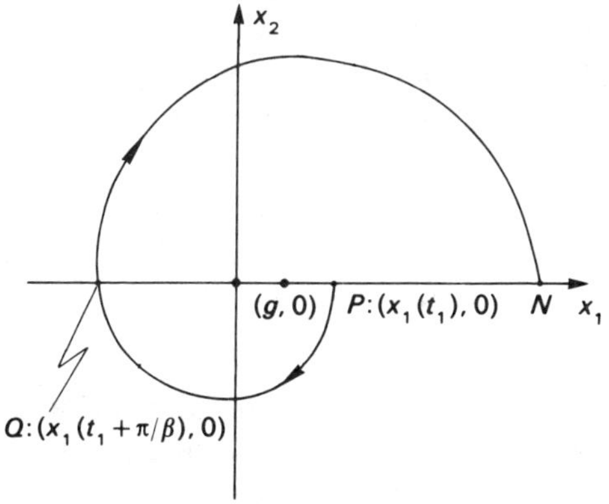

Fig. 4.34. The phase plane of the system defined by (4.122) and (4.123) with $x_1 = x$ and $x_2 = \dot{x}$. The trajectory evolving from $(x_1(t_1), 0)$ is shown. For $x_2 \leqslant 0$, this trajectory spirals about $(0, 0)$; for $x_2 > 0$ it spirals about $(g, 0)$.

For $t > t_1 + \pi/\beta$ the phase point moves into the upper half plane in Fig. 4.34 and its motion is governed by (4.122). It consequently executes part of a spiral centred on $(g, 0)$, until at $t = t_1 + 2\pi/\beta$ the positive x_1-axis is encountered at N. Equation (4.126) gives

$$x_1\left(t_1 + \frac{\pi}{\beta}\right) - g = \bar{R}\exp\left[-k\left(t_1 + \frac{\pi}{\beta}\right)\right]\cos(\beta t_1 + \pi + \bar{\theta})$$

and

$$x_1\left(t_1 + \frac{2\pi}{\beta}\right) - g = -\bar{R}\exp\left[-k\left(t_1 + \frac{\pi}{\beta}\right)\right]\rho\cos(\beta t_1 + \pi + \bar{\theta}). \tag{4.130}$$

Thus, using (4.129) we have

$$x_1\left(t_1 + \frac{2\pi}{\beta}\right) - g = -[-\rho x_1(t_1) - g]\rho. \tag{4.131}$$

If a closed trajectory exists, there must be a value of $x_1(t_1)$ such that N coincides with P. This will be so provided there is a positive solution to (4.131) with $x_1(t_1 + 2\pi/\beta) = x_1(t_1)$. It is easily seen that there is only one such solution, namely

$$x_1(t_1) = \frac{g}{1-\rho} = x_1^{(c)}. \tag{4.132}$$

Consequently there is a single closed trajectory.

We may confirm that the closed orbit is an attracting limit cycle by arguing as follows. Consider the trajectory starting from an arbitrary point in the x_1x_2-plane and let $\bar{x}_1 \neq x_1^{(c)}$ be the x_1-coordinate of its first intersection with the positive x_1-axis. Equation (4.131) then implies that the x_1-coordinates of intersections with the positive x_1-axis are given by the sequence

$$\{\bar{x}_1,\ g(1+\rho) + \rho^2\bar{x}_1,\ g(1+\rho+\rho^2+\rho^3) + \rho^4\bar{x}_1, \ldots,$$
$$g(1+\rho+\rho^2+\ldots+\rho^{2m-1}) + \rho^{2m}\bar{x}_1, \ldots\}, \tag{4.133}$$

where m is the number of complete revolutions of the trajectory about the origin. Since $\rho < 1$ the limit of the sequence (4.133) exists and is equal to the sum of the geometric series

$$g\sum_{i=0}^{\infty}\rho^i = \frac{g}{1-\rho} = x_1^{(c)}. \tag{4.134}$$

Thus all trajectories 'home in' on the closed one and we conclude that the system has a unique, attracting limit cycle.

Example 4.5.2. Consider the electrical circuit shown in Fig. 4.35 (Andronov and Chaikin, 1949). There are two unfamiliar circuit elements involved here. The *triode valve*, T, consists of an evacuated glass envelope with three electrodes which are connected to nodes 1, 2 and 3 of the network. When a potential difference is applied between node 2 and node 1 a current flows through the triode however the magnitude of this current is a function, f say, of the potential difference between node 3 and node 1. This function is called the mutual characteristic of T. Taking currents j and i as shown to satisfy Kirchhoff's current law, we have

$$i+j=f(v_{31}). \tag{4.135}$$

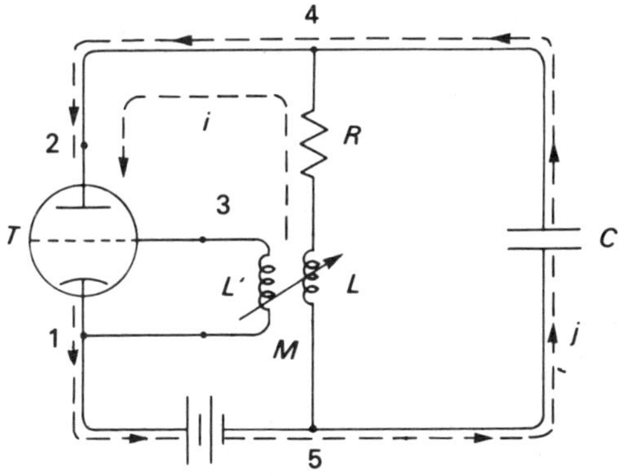

Fig. 4.35. An electrical circuit which provides a realization of the dynamical equations (4.122) and (4.123).

The origin of v_{31} involves the second new circuit element. This is the *mutual inductance*, M. This consists of two coils L and L' wound together; a change of current, i, in L produces a potential difference of $M\mathrm{d}i/\mathrm{d}t$ across L', i.e.

$$v_{31} = M\mathrm{d}i/\mathrm{d}t, \tag{4.136}$$

where M is a positive constant. Find the dynamical equation governing the current through the resistor R.

Solution. If v_L, v_R, v_C are the potential differences across the corresponding elements, in the direction of the currents i and j, then Kirchhoff's voltage law and the properties of L, C and R give

$$v_L + v_R = v_C; \quad \mathrm{L}\frac{\mathrm{d}i}{\mathrm{d}t} = v_L; \quad C\frac{\mathrm{d}v_C}{\mathrm{d}t} = j; \quad v_R = iR. \tag{4.137}$$

Equations (4.135) and (4.136) imply

$$i + j = f\left(M\frac{\mathrm{d}i}{\mathrm{d}t}\right),$$

while (4.137) gives

$$\frac{\mathrm{d}v_C}{\mathrm{d}t} = \frac{j}{C} = \frac{\mathrm{d}v_L}{\mathrm{d}t} + \frac{\mathrm{d}v_R}{\mathrm{d}t} = L\frac{\mathrm{d}^2 i}{\mathrm{d}t^2} + R\frac{\mathrm{d}i}{\mathrm{d}t}.$$

Thus

$$\frac{\mathrm{d}^2 i}{\mathrm{d}t^2} + \frac{R}{L}\frac{\mathrm{d}i}{\mathrm{d}t} + \frac{i}{LC} = \frac{1}{L}f\left(M\frac{\mathrm{d}i}{\mathrm{d}t}\right). \tag{4.138}$$ □

Now suppose $f(v)$ is simply a step function, i.e.

$$f(v) = \begin{cases} 0, & v \leqslant 0 \\ I_0/C, & v > 0 \end{cases},$$

then, defining $2k = R/L$ and $\omega_0^2 = 1/LC$, we have

$$\frac{\mathrm{d}^2 i}{\mathrm{d}t^2} + 2k\frac{\mathrm{d}i}{\mathrm{d}t} + \omega_0^2 i = \begin{cases} \omega_0^2 I_0, & \mathrm{d}i/\mathrm{d}t > 0 \\ 0, & \mathrm{d}i/\mathrm{d}t \leqslant 0 \end{cases}. \tag{4.139}$$

We conclude therefore that the triode valve with a discontinuous mutual characteristic can provide a realization of the dynamical equations (4.122) and (4.123).

Exercises

Section 4.1

1. For the harmonic oscillator equations

$$\dot{x}_1 = x_2, \qquad \dot{x}_2 = -\omega_0^2 x_1 - 2kx_2,$$

write down the canonical system $\dot{\boldsymbol{y}} = \boldsymbol{J}\boldsymbol{y}$ when
(a) $k = 0$, (b) $0 < k < \omega_0$, (c) $k = \omega_0$, (d) $k > \omega_0$,

and its solutions. Without finding the transformation matrix $\boldsymbol{M}$ explicitly, show that the solutions in the x_1, x_2 variables take the form given in Equations (4.5)–(4.8).

2. Sketch the phase portrait for the canonical system $\dot{\boldsymbol{y}} = \boldsymbol{J}\boldsymbol{y}$ corresponding to the critically damped harmonic oscillator

$$\dot{x}_1 = x_2, \qquad \dot{x}_2 = -k^2 x_1 - 2kx_2.$$

Show that $\boldsymbol{x} = \boldsymbol{M}\boldsymbol{y}$ where

$$\boldsymbol{M} = \begin{bmatrix} 1 & 0 \\ -k & 1 \end{bmatrix}$$

and write down the principal directions at the origin. Use these directions and the method of isoclines to sketch the phase portrait in the x_1x_2-plane. Is this the same as Fig. 4.5(a)?

3. Consider an overdamped harmonic oscillator

$$\dot{x}_1 = x_2, \qquad \dot{x}_2 = -\omega_0^2 x_1 - 2kx_2, \qquad k > \omega_0.$$

Write down the canonical system $\dot{\boldsymbol{y}} = \boldsymbol{J}\boldsymbol{y}$. Find the principal directions at the origin and investigate their behaviour as $k \to \infty$. Show that for large k the trajectories in the x_1x_2-plane are essentially vertical except near to the x_1-axis.

4. Consider a circuit with three elements, an inductor L, capacitor C and a linear resistor R connected in parallel between two nodes. Show that current through the inductor satisfies a damped harmonic oscillator equation. Identify the damping k and natural frequency ω_0. Assuming L and R are fixed, find a condition on C for the inductor current to execute damped oscillations.

5. Consider the circuit shown below.

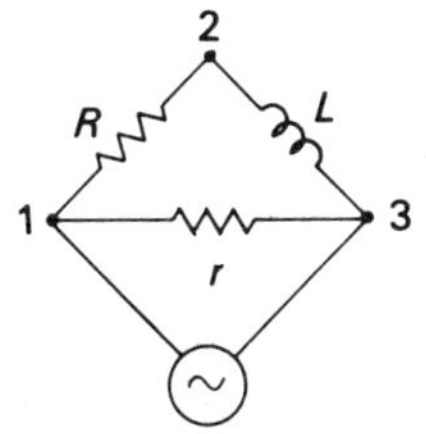

Suppose that

$$v_{13} = E(t) = \begin{cases} 0, & t \leqslant 0 \\ E_0, & t > 0. \end{cases}$$

Show that the current flowing through the resistor R at time $t > 0$ is

$$j_R = \frac{E_0}{R}(1 - e^{-Rt/L}).$$

6. An electrical circuit consists of a capacitor C and a linear resistor R joined in parallel between two nodes. Prove that the potential difference across the resistor decays exponentially to zero regardless of its initial value.

7. A model for price adjustment in relation to stock level is given as follows. The rate of change of stock (q) is assumed to be proportional to the difference between supply (s) and demand (u), i.e.

$$\dot{q} = k(s - u), \quad k > 0.$$

The rate of change of price (p) is taken to be proportional to the amount by which stock falls short of a given level q_0 and so

$$\dot{p} = -m(q - q_0), \quad m > 0.$$

If both s and u are assumed to be affine functions of p, find a second order differential equation for p. Show that if $u > s$ when $p = 1$ and $\frac{ds}{dp} > \frac{du}{dp}$, then the price oscillates with time.

8. A cell population consists of 2-chromosome and 4-chromosome cells. The dynamics of the population are modelled by

$$\dot{D} = (\lambda - \mu)D, \qquad \dot{U} = \mu D + \nu U,$$

where D and U are the numbers of 2- and 4-chromosome cells, respectively. Assume that $D = D_0$ and $U = U_0$ at $t = 0$ and find the proportion of 2-chromosome cells present in the population as a function of t. Show that this tends to a saturation level independent of D_0 and U_0 providing $\lambda > \mu + \nu$.

9. The motion of a particle P moving in the plane with coordinates x

and y is governed by the differential equations

$$\ddot{x} = -\omega^2 x, \qquad \ddot{y} = -y.$$

Plot the parametrized curve $(x(t), y(t))$ in the xy-plane when $x(0) = 0$, $\dot{x}(0) = 1$, $y(0) = 1$, $\dot{y}(0) = 0$ for $\omega = 1, 2, 3$. [These dynamical equations are realized in the *biharmonic oscillator* (Arnold, 1978).]

10. Three springs (each of natural length l and spring constant k) and two masses, m, are arranged as shown below on a smooth horizontal table. The ends A and B are held fixed and the masses are displaced along the line of the springs and released.

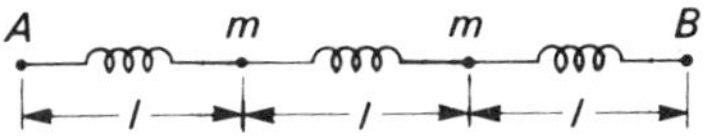

Let x_1 and x_2, respectively, be the displacements (measured in the same sense) of the masses from their equilibrium positions. Show that the equations of motion of the masses are

$$m\ddot{x}_1 = k(-2x_1 + x_2), \qquad m\ddot{x}_2 = k(x_1 - 2x_2).$$

Write these equations in the matrix form $\ddot{\boldsymbol{x}} = \boldsymbol{A}\boldsymbol{x}$, where $\boldsymbol{x} = \begin{bmatrix} x_1 \\ x_2 \end{bmatrix}$, and find a linear change of variables, $\boldsymbol{x} = \boldsymbol{M}\boldsymbol{y}$, such that $\ddot{\boldsymbol{y}} = \boldsymbol{D}\boldsymbol{y}$ with $\boldsymbol{D}$ diagonal. Hence, find normal coordinates and describe the normal modes of oscillation of the two masses.

Section 4.2

11. In a simple model of a national economy, $\dot{I} = I - \alpha C$, $\dot{C} = \beta(I - C - G)$, where I is the national income, C is the rate of consumer spending and G is the rate of government expenditure. The model is restricted to its natural domain $I \geqslant 0, C \geqslant 0, G \geqslant 0$ and the constants α and β satisfy $1 < \alpha < \infty, 1 \leqslant \beta < \infty$.

(a) Show that if the rate of government expenditure $G = G_0$, a constant, then there is an equilibrium state. Classify the equilibrium state when $\beta = 1$ and show that then the economy oscillates.

(b) Assume government expenditure is related to the national income by the rule $G = G_0 + kI$, where $k > 0$. Find the upper bound A

on k for which an equilibrium state exists in the natural domain of this model. Describe both the position and the behaviour of this state for $\beta > 1$, as k tends to the critical value A.

12. Suppose $\boldsymbol{U}(s)$ and $\boldsymbol{V}(s)$ are matrices with elements $u_{ij}(s)$ and $v_{ij}(s)$ $(i, j = 1, \ldots, n)$. Verify the relation

$$\frac{\mathrm{d}}{\mathrm{d}s}(\boldsymbol{UV}) = \frac{\mathrm{d}\boldsymbol{U}}{\mathrm{d}s}\boldsymbol{V} + \boldsymbol{U}\frac{\mathrm{d}\boldsymbol{V}}{\mathrm{d}s} = \dot{\boldsymbol{U}}(s)\boldsymbol{V}(s) + \boldsymbol{U}(s)\dot{\boldsymbol{V}}(s)$$

and show that

$$\int_0^t \boldsymbol{U}(s)\dot{\boldsymbol{V}}(s)\mathrm{d}s = [\boldsymbol{U}(s)\boldsymbol{V}(s)]_0^t - \int_0^t \dot{\boldsymbol{U}}(s)\boldsymbol{V}(s)\mathrm{d}s. \qquad (1)$$

Let

$$\boldsymbol{P} = \int_0^t \mathrm{e}^{-\boldsymbol{A}s}\begin{bmatrix} 0 \\ \cos(\omega s) \end{bmatrix}\mathrm{d}s$$

and

$$\boldsymbol{Q} = \int_0^t \mathrm{e}^{-\boldsymbol{A}s}\begin{bmatrix} 0 \\ \sin(\omega s) \end{bmatrix}\mathrm{d}s.$$

Use (1) to obtain:

(a) $\omega\boldsymbol{P} = \mathrm{e}^{-\boldsymbol{A}t}\begin{bmatrix} 0 \\ \sin(\omega t) \end{bmatrix} + \boldsymbol{AQ}$;

(b) $\omega\boldsymbol{Q} = \left\{-\mathrm{e}^{-\boldsymbol{A}s}\begin{bmatrix} 0 \\ \cos(\omega s) \end{bmatrix}\right\}_0^t - \boldsymbol{AP}.$

Hence show that

$$(\omega^2\boldsymbol{I} + \boldsymbol{A}^2)\boldsymbol{P} = \boldsymbol{A}\begin{bmatrix} 0 \\ 1 \end{bmatrix} + \mathrm{e}^{-\boldsymbol{A}t}\left\{\begin{bmatrix} 0 \\ \omega\sin(\omega t) \end{bmatrix} - \boldsymbol{A}\begin{bmatrix} 0 \\ \cos(\omega t) \end{bmatrix}\right\}$$

and obtain Equation (4.59).

13. Consider the electrical circuit shown below, where $u(t)$ is an 'input' voltage applied at A and $y(t)$ is the corresponding 'output' voltage measured across the smaller capacitor at B.

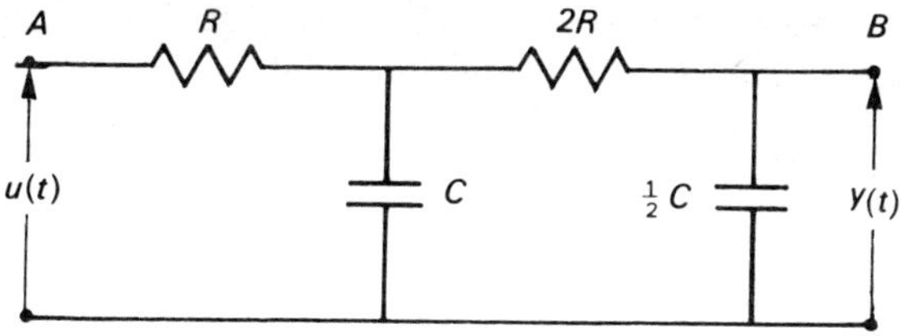

Obtain the dynamical equation

$$R^2C^2\,\ddot{y}(t) + \tfrac{5}{2}RC\dot{y}(t) + y(t) = u(t)$$

and derive the equivalent first-order system

$$\frac{\mathrm{d}x_1}{\mathrm{d}\tau} = x_2, \qquad \frac{\mathrm{d}x_2}{\mathrm{d}\tau} = -x_1 - \tfrac{5}{2}x_2 + u(RC\tau)$$

where $RC\tau = t$ and $x_1 = y$. Hence show that the steady-state output of the circuit is given by

$$y_s(t) = \frac{2}{3RC}\int_{t_0}^{t} \left[\mathrm{e}^{(s-t)/2RC} - \mathrm{e}^{2(s-t)/RC}\right]u(s)\mathrm{d}s$$

for any initial values of $y(t_0)$ and $\dot{y}(t_0)$.

14. Consider the electrical circuit shown below.

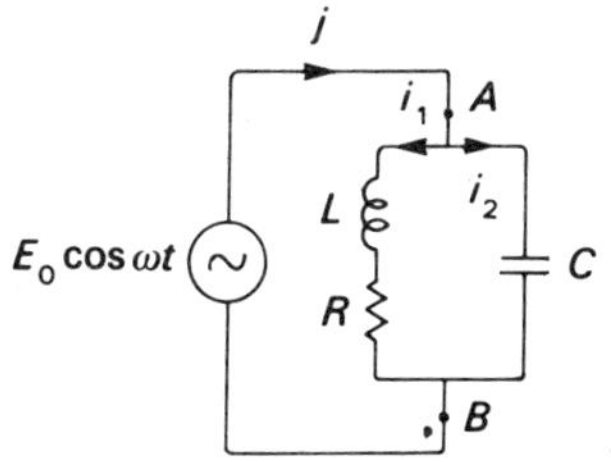

Obtain differential equations for the currents i_1 and i_2 and hence find their steady-state values. Calculate the amplitude, j_0, of the current j. Show that the impedance Z $(= E_0/j_0)$ of the LCR circuit between A and B is given by

$$Z = \frac{L/C}{[R^2 + (\omega L - 1/\omega C)]^{1/2}},$$

where R is small compared with ωL. Sketch Z as a function of ω and obtain the resonant frequency of the circuit.

Section 4.3

15. Investigate the nature of the fixed points of the competing species model

$$\dot{x}_1 = (2 - x_1 - 2x_2)x_1, \qquad \dot{x}_2 = (2 - 2x_1 - x_2)x_2$$

and indicate their position, together with the $\dot{x}_1 = 0$, $\dot{x}_2 = 0$ isoclines, in the x_1x_2-plane. Find the principal directions at the fixed points, sketch the phase portrait and interpret it in terms of species behaviour.

16. Examine the behaviour of the fixed points of the competing species model

$$\dot{x}_1 = (1 - x_1 - x_2)x_1, \qquad \dot{x}_2 = (\nu - x_2 - 4\nu^2 x_1)x_2, \qquad x_1, x_2 > 0,$$

as ν varies through positive values. Show that changes in the number and the nature of the fixed points occur at $\nu = \frac{1}{4}$ and $\nu = 1$. Sketch typical phase portraits for ν in the intervals $(0, \frac{1}{4})$, $(\frac{1}{4}, 1)$ and $(1, \infty)$.

17. Consider the prey–predator equations with 'logistic' corrections

$$\dot{x}_1 = x_1(1 - x_2 - \alpha x_1), \qquad \dot{x}_2 = -x_2(1 - x_1 + \alpha x_2),$$

where $0 \leqslant \alpha < 1$. Show that, at the non-trivial fixed point, the centre that exists for $\alpha = 0$ changes into a stable focus for $0 < \alpha < 1$.
Sketch the phase portrait.

18. Let $(x_1(t), x_2(t))$ be a periodic solution of the prey–predator equations

$$\dot{x}_1 = x_1(a - bx_2), \qquad \dot{x}_2 = -x_2(c - dx_1).$$

Define the average value, $\bar{x}_i$, of x_i by

$$\bar{x}_i = \frac{1}{T}\int_0^T x_i(t)\mathrm{d}t$$

where T is the period of the solution. Show that $\bar{x}_1 = c/d$ and $\bar{x}_2 = a/b$.

Suppose the dynamical equations are modified by the addition of 'harvesting' terms, $-\varepsilon x_i (\varepsilon > 0)$, to $\dot{x}_i$ for $i = 1, 2$. Such terms correspond, for example, to the effects of fishing on fish populations or chemical sprays on insect populations. What effect do these harvesting terms have on the average populations, $\bar{x}_1$ and $\bar{x}_2$?

19. Show that the fixed point $(1, J^{-1})$ of the Holling–Tanner model (4.90) is stable if the $\dot{y}_2 = 0$ isocline intersects the parabola $\dot{y}_1 = 0$ to the right of its peak. Hence show that if a phase portrait for this model contains only one limit cycle, which is stable, then the $\dot{y}_2 = 0$ isocline must intersect the parabola to the left of its peak.

20. An age-dependent population model is given by

$$\dot{P} = -\mu(P)P + B, \qquad \dot{B} = [\gamma - \mu(P)]B, \quad \gamma > 0,$$

where P is the total population and B is the birth rate. Prove that $B = \gamma P$ is a union of trajectories for all choices of the function $\mu(P)$. Investigate the phase portrait when $\mu(P) = b + cP$ where $b < 0$ and $c > 0$. Show that for all positive initial values of the variables, both population and birth rate stabilize at non-zero values.

21. Find a first integral for the general epidemic model

$$\dot{x} = -2xy, \qquad \dot{y} = 2xy - y,$$

where x is the number of susceptibles and y is the number of infectives, suitably scaled. Hence, or otherwise, sketch the phase portrait in the region $x, y \geqslant 0$. Show that the number of infectives reaches a peak of

$$c_0 - \tfrac{1}{2}(1 + \ln 2)$$

when $x = \frac{1}{2}$, where c_0 is the total number of susceptibles and infectives when $x = 1$. How does the epidemic evolve?

22. In Exercise 4.21, suppose the stock of susceptibles is being added to at a constant rate so that

$$\dot{x} = -2xy + 1, \qquad \dot{y} = 2xy - y.$$

Show that the new system has a stable fixed point in the region $x, y > 0$. What are the implications for the development of this epidemic?

23. The system

$$\dot{S} = -rIS, \qquad \dot{I} = rIS - \gamma I, \qquad R = 1 - S - I$$

$(r, \gamma > 0)$ models how a disease, which confers permanent immunity, spreads through a population. Let S, I and R be the fractions of the population which are, respectively, susceptible, infected and immune. Define $\sigma = r/\gamma$ and assume initial values $S = S_0$, $I = I_0$ and $R = 0$. Prove that:

(a) if $\sigma S_0 \leqslant 1$, then $I(t)$ decreases to zero;

(b) if $\sigma S_0 > 1$, then $I(t)$ increases to a maximum value of $1 - (1 + \ln(\sigma S_0))/\sigma$ and then decreases to zero.

Show that in both (a) and (b) the population $S(t)$ approaches S_L as $t \to \infty$, where S_L is the unique root of the equation $S_L = (1/\sigma)\ln(S_L/S_0) + 1$ in the interval $(0, 1/\sigma)$.

24. A simple model of the molecular control mechanism in cells involves the quantity X of messenger ribonucleic acid and the quantity Y of a related enzyme. The dynamical equations are given by

$$\dot{X} = \frac{a}{A + kY} - b, \qquad \dot{Y} = \alpha X - \beta, \qquad a > bA,$$

where A, k, a, b, α, β are positive constants. Prove that both X and Y exhibit persistent oscillations in time.

25. Investigate the fixed points of the equations of motion

$$\dot{x}_1 = x_2, \qquad \dot{x}_2 = -\omega_0^2 \sin x_1$$

of the simple pendulum where $\omega_0^2 = g/l$. Here, g is the acceleration due to gravity and l is the length of the pendulum. Find a first integral and sketch the phase portrait. Suppose a damping term $-2kx_2$, $k > 0$, is added to $\dot{x}_2$. Find the nature of the fixed points of the damped system when k is small and sketch its phase portrait.

Interpret both of the above phase portraits in terms of the motions of the pendulum.

26. The behaviour of a simple disc dynamo is governed by the system

$$\dot{x} = -\mu x + xy, \qquad \dot{y} = 1 - \nu y - x^2; \qquad \mu, \nu > 0,$$

where x is the output current of the dynamo and y is the angular velocity of the rotating disc. Prove that for $\mu v > 1$ there is one stable fixed point A at $(0, v^{-1})$ but for $\mu v < 1$, A becomes a saddle point and stable fixed points occur at $(\pm\sqrt{(1-\mu v)}, \mu)$.

Section 4.4

27. Verify that the system

$$\dot{x}_1 = x_2, \qquad \dot{x}_2 = -x_1 - \varepsilon(x_1^2 - 1)x_2, \qquad \varepsilon \geqslant 0, \tag{1}$$

is equivalent to:

(a) the Van der Pol equation $\varepsilon > 0$;

(b) the undamped harmonic oscillator when $\varepsilon = 0$.

Obtain a system equivalent to (1) in polar coordinates, where $x_1 = r\cos\theta$ and $x_2 = r\sin\theta$, and derive an expression for $dr/d\theta$.

Assume ε is small compared with unity and suppose the trajectory passing through $(r, \theta) = (r_0, 0)$ evolves to $(r_1, -2\pi)$. Show that

$$\Delta r = r_1 - r_0 = \int_0^{2\pi} \varepsilon r_0 \sin^2\theta(1 - r_0^2\cos^2\theta)\,d\theta$$

to first order in ε. Evaluate this integral and explain why the result implies the existence of a stable limit cycle with approximate radius 2.

28. (a) Show that the Van der Pol equation is obtained by differentiating the Rayleigh equation,

$$\ddot{x} + \varepsilon(\tfrac{1}{3}\dot{x}^3 - \dot{x}) + x = 0,$$

with respect to time and setting $y = \dot{x}$. Show also that the two equations can be represented by the same first-order system.

(b) Consider the Liénard representation

$$\dot{x}_1 = x_2 - \varepsilon(\tfrac{1}{3}x_1^3 - x_1), \qquad \dot{x}_2 = -x_1$$

of the Van der Pol equation. Show that the dependence of the characteristic on the parameter ε can be removed by rescaling the variable x_2 and letting $x_2 = \varepsilon\omega$. Sketch the phase portrait of the Van der Pol oscillator in the $x_1\omega$-plane as $\varepsilon \to \infty$.

29. Let a resistor R, with a current-voltage characteristic of $j = v^3 - v$, be connected to an inductor L to form a single loop. With the notation in the figure, show that the dynamics of the circuit are

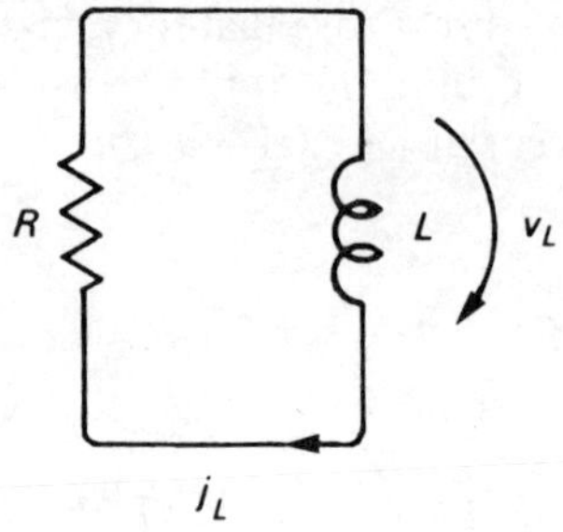

given by

$$L\frac{\mathrm{d}j_L}{\mathrm{d}t} = v_L$$

for $j_L = v_L - v_L^3$.

Regularize this circuit by introducing a small capacitor in an appropriate way and deduce that the regularized dynamical equations can oscillate.

Section 4.5

30. A model for a population which becomes susceptible to epidemics is constructed as follows. The population is originally governed by

$$\dot{p} = ap - bp^2 \tag{1}$$

and grows to a certain value $Q < a/b$. At this population the epidemic strikes and the population is governed by

$$\dot{p} = Ap - Bp^2 \tag{2}$$

where $Q > A/B$. The population falls to the value q, where $A/B < q < a/b$, at which point the epidemic ceases and the population is again controlled by (1) and so on. Sketch curves in the p–t plane to illustrate the fluctuations in population with time. Show the time T_1 for the population to increase from q to Q is given by

$$T_1 = \frac{1}{a}\ln\left[\frac{Q(a-bq)}{q(a-bQ)}\right].$$

Find the time T_2 taken for the population to fall from Q to q under influence of (2) and deduce the period of a typical population cycle.

31. A model of a trade cycle is given by

$$\ddot{Y} - \phi(\dot{Y}) + k\dot{Y} + Y = 0, \qquad 0 < k < 2,$$

where Y is the output. The function ϕ satisfies $\phi(0) = 0$,

$$\phi(x) \to L \quad \text{as } x \to \infty \quad \text{and} \quad \phi(x) \to -M \quad \text{as } x \to -\infty,$$

where M is the scrapping rate of existing stock and L is the net amount of capital-goods trade over and above M. If the function ϕ is idealized to be

$$\phi(0) = 0; \qquad \phi(y) = L, \quad y > 0; \qquad \phi(y) = -M, \quad y < 0,$$

sketch the phase portrait of the differential equation. Show that the output $Y(t)$, regardless of its initial value, eventually oscillates with a fixed amplitude and period.

32. A model of economic growth is given by $\ddot{Y} + 2\dot{Y} + Y = G(t)$, where $Y(t)$ is the output and $G(t)$ is government spending. The function $G(t)$ has the form $G(t) = 0, 0 \leqslant t < 1$ and $G(t) = G_0, t \geqslant 1$.

Show that there exists an output curve $Y_1(t)$ with the following features:

(a) $Y_1(0) = 0$;

(b) $Y_1(t)$ increases for $t \in [0, 1]$ to a maximum value G_0 where $\dot{Y}_1(1) = 0$;

(c) $Y_1(t) = G_0, \quad t \geqslant 1$.

Investigate the long-term effect on the output $Y(t)$ if, with the same initial conditions ((a) and (b)) as for $Y_1(t)$, the onset of government spending occurs at a time later than $t = 1$.

33. Consider a block of mass m whose motion on a horizontal conveyor belt is constrained by a light spring as shown below.

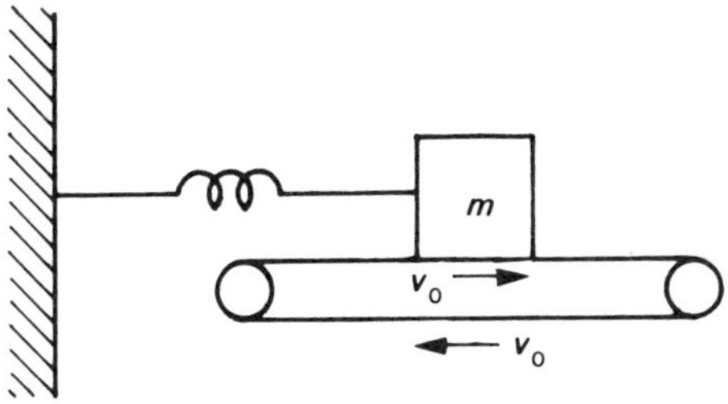

Suppose the belt is driven at a constant speed v_0 and let $x(t)$ be the extension of the spring at time t. The frictional force F exerted on the block by the belt is taken to be

$$F = \begin{cases} F_0, & \dot{x} < v_0 \\ -F_0, & \dot{x} > v_0, \end{cases}$$

i.e. 'dry' or 'Coulomb' friction.

Show that the equation of motion of the block is

$$m\ddot{x} + kx = -F$$

and sketch the phase portrait in the $x\dot{x}$-plane. Describe the possible motions of the block.

CHAPTER FIVE

Advanced techniques and applications

In this final chapter we present a selection of theoretical items and recent models which indicate how the basic ideas of the earlier chapters can be developed. Our aim is to fill some theoretical gaps and to stimulate interest in current applications.

5.1 The Liénard equation

In this section we prove that the system

$$\dot{x}_1 = x_2 - x_1^3 + x_1, \qquad \dot{x}_2 = -x_1 \tag{5.1}$$

has a unique attracting limit cycle. The method of proof also works for systems of the form

$$\dot{x}_1 = x_2 - G(x_1), \qquad \dot{x}_2 = -h(x_1), \tag{5.2}$$

for which $G(x_1)$ behaves in a similar way (see Theorem 5.1.2) to the cubic characteristic $x_1^3 - x_1$ and $h(x_1)$ is odd. Such systems correspond to the second-order equations

$$\ddot{x} + g(x)\dot{x} + h(x) = 0, \qquad g(x) = \mathrm{d}G(x)/\mathrm{d}x, \tag{5.3}$$

which are called *Liénard equations*. For example, the Van der Pol equation

$$\ddot{x} + \varepsilon(x^2 - 1)\dot{x} + x = 0, \qquad \varepsilon > 0, \tag{5.4}$$

has the cubic characteristic $G(x_1) = \varepsilon(\frac{1}{3}x_1^3 - x_1)$; and hence a unique attracting limit cycle, for any value of $\varepsilon > 0$.

Theorem 5.1.1. The phase portrait of the system

$$\dot{x}_1 = x_2 - x_1^3 + x_1, \qquad \dot{x}_2 = -x_1 \tag{5.5}$$

consists of a unique attracting limit cycle surrounding an unstable focus at the origin.

Proof. The proof is carried out in two parts.

(a) We prove that all trajectories spiral around the origin. Consider the trajectory through the point $\mathbf{x}_0$ on the positive x_2-axis (labelled P in Fig. 5.1). At P, $\dot{x}_1 > 0$ and $\dot{x}_2 = 0$ and so the trajectory $\phi_t(\mathbf{x}_0)$ (ϕ is taken to be the flow of system (5.5)) enters the region $x_1 > 0$ for $t > 0$. For $x_1 > 0$, $\dot{x}_2 < 0$ and so the trajectory curves downwards to some point $Q(= \phi_{t_Q}(\mathbf{x}_0))$ on the $\dot{x}_1 = 0$ isocline. As t increases through t_Q the trajectory enters a region for which $\dot{x}_1 < 0$ and $\dot{x}_2 < 0$. Hence the trajectory intersects the $\mathrm{d}x_2/\mathrm{d}x_1 = 1$ isocline at the point $R(= \phi_{t_R}(\mathbf{x}_0))$. For $t > t_R$ the trajectory enters the region $x_2 < x_1^3 - 2x_1$ where $\mathrm{d}x_2/\mathrm{d}x_1 < 1$. Thus the straight line of slope 1 through R bounds the trajectory below. In the region $x_2 < x_1^3 - 2x_1$ both $\dot{x}_1$ and $\dot{x}_2$ are negative and so the trajectory through P meets the negative x_2-axis at S.

Observe that the system (5.5) is unchanged under the reflection in the origin $(x_1, x_2) \to (-x_1, -x_2)$. Thus the trajectory through S is just the reflection in the origin of the trajectory through the point S^* on the positive x_2-axis where $OS^* = OS$ (see Fig. 5.1). We conclude that the trajectory through a point P on the positive x_2-axis spirals clockwise around the origin to intersect the positive x_2-axis repeatedly.

(b) The second part of the proof consists of showing that there is one and only one trajectory which closes to form a stable limit cycle.

Suppose A is a point on the positive x_2-axis. Let A' be the first point of return of the trajectory through A to the positive x_2-axis. Let $OA' = a'$ and define the function $f: \mathsf{R}^+ \to \mathsf{R}$ by $f(a) = a' - a$. Note that the trajectory through A will be closed (periodic) if and only if $f(a) = 0$, that is when $A = A'$. The map f just gives the signed change in the radial distance r for a complete revolution.

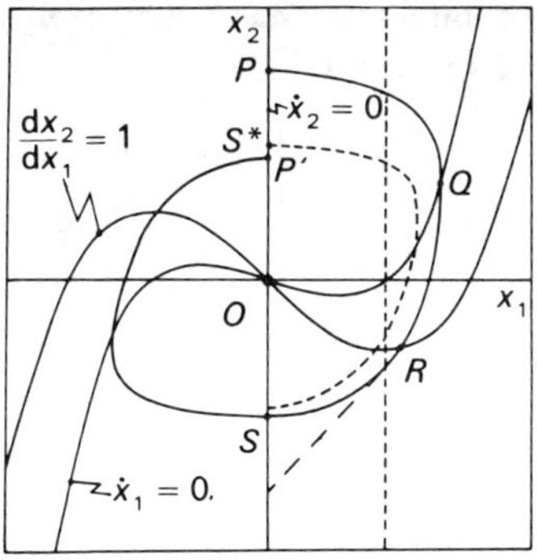

Fig. 5.1. The isoclines $\dot{x}_1 = 0$, $\dot{x}_2 = 0$, $dx_2/dx_1 = 1$ and the trajectories through P and S^* where $OS^* = OS$.

If r denotes radial distance in the x_1x_2-plane then $r\dot{r} = x_1\dot{x}_1 + x_2\dot{x}_2$ and for system (5.1)

$$r\dot{r} = x_1^2(1 - x_1^2). \tag{5.6}$$

Let

$$\Delta(P_0, P_1) = \int_{t_0}^{t_1} r\dot{r}\,dt = \int_{t_0}^{t_1} x_1^2(1 - x_1^2)dt. \tag{5.7}$$

Then $\Delta(P_0, P_1)$ is proportional to the change in $r^2(t)$ on a solution curve $\mathbf{x}(t)\{= (x_1(t), x_2(t))\}$ between P_0 when $t = t_0$ and P_1 when $t = t_1$. Let $\mathbf{x}(t)$, $\mathbf{x}^*(t)$ be the solutions corresponding to trajectories through the points A and A_1 where $OA < OA_1$ (see Fig. 5.2). Suppose

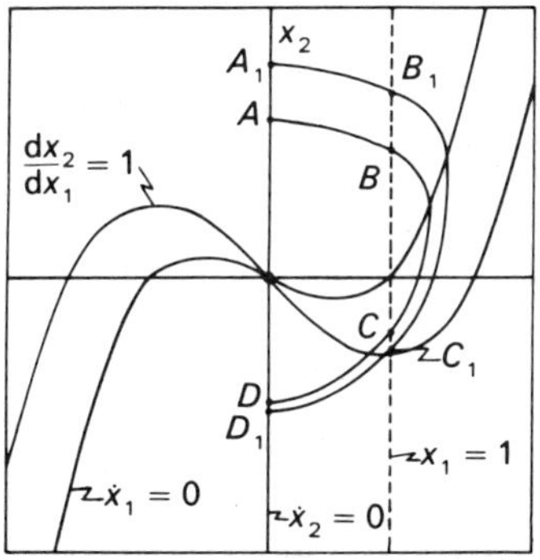

Fig. 5.2. Trajectories passing through A and A_1 on the positive x_2-axis and their points of intersection with the line $x_1 = 1$.

the two trajectories intersect the negative x_2-axis at D and D_1 respectively. We will show that

$$\Delta(A_1,D_1) < \Delta(A,D). \tag{5.8}$$

The solution curves are split into three segments (see Fig. 5.2):

(a) AB and $A_1 B_1$

The change

$$\Delta(A, B) = \int_{x_1=0}^{x_1=1} x_1(t)^2[1 - x_1(t)^2]\mathrm{d}t = \int_0^1 \frac{x_1^2(1-x_1^2)}{x_2 - x_1^3 + x_1}\mathrm{d}x_1, \tag{5.9}$$

using $\mathrm{d}x_1/\mathrm{d}t = x_2 - x_1^3 + x_1$. Similarly

$$\Delta(A_1, B_1) = \int_0^1 \frac{x_1^2(1-x_1^2)}{x_2^* - x_1^3 + x_1}\mathrm{d}x_1. \tag{5.10}$$

However for $x_1 \in [0, 1], x_2^* > x_2 > x_1^3 - x_1$ where (x_1, x_2) and (x_1, x_2^*) are points of the curves AB and $A_1 B_1$ respectively and so

$$\Delta(A_1, B_1) < \Delta(A, B). \tag{5.11}$$

(b) BC and B_1C_1

The segments BC and B_1C_1 can be parametrized by x_2 as functions $x_1 = x_1(x_2)$ and $x_1 = x_1^*(x_2)$ (see Fig. 5.3). With the notation in Fig. 5.3,

$$\Delta(B, C) = \int_{x_2=b}^{x_2=c} x_1(x_2)^2[1 - x_1(x_2)^2]\mathrm{d}t$$
$$= -\int_c^b x_1(x_2)[x_1(x_2)^2 - 1]\mathrm{d}x_2, \tag{5.12}$$

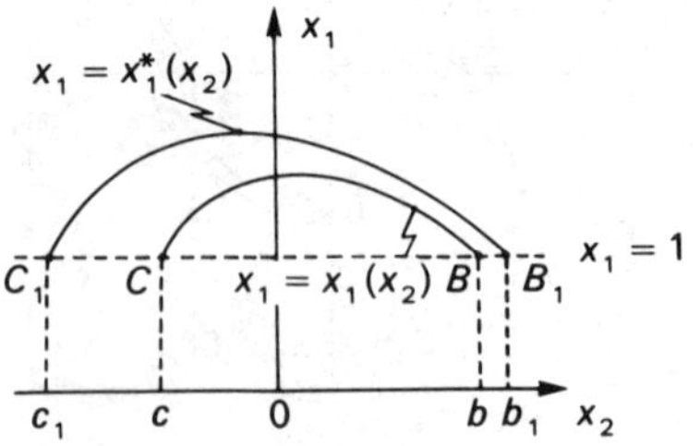

Fig. 5.3. The segments BC and B_1C_1 of the two solution curves through A and A_1.

using $\dot{x}_2 = -x_1$. The change $\Delta(B_1, C_1)$ is given by the same expression as (5.12) with $x_1(x_2)$ replaced by $x_1^*(x_2)$ together with an interval of integration $[c_1, b_1]$. For any fixed $x_2 \in [c, b]$, $x_1^*(x_2) > x_1(x_2) \geqslant 1$ and so

$$x_1^*(x_2)[x_1^*(x_2)^2 - 1] \geqslant x_1(x_2)[x_1(x_2)^2 - 1] \geqslant 0. \tag{5.13}$$

Thus noting the minus sign in (5.12)

$$\Delta(B_1 C_1) < \Delta(B, C). \tag{5.14}$$

(c) CD and $C_1 D_1$
Similar considerations to those in case (a) imply

$$\Delta(C_1, D_1) < \Delta(C, D). \tag{5.15}$$

Adding the three inequalities obtained above we conclude

$$\Delta(A_1, D_1) < \Delta(A, D). \tag{5.16}$$

The symmetry of the Liénard equation (5.5) allows us to deduce that in the region $x_1 < 0$, the radial change Δ in the 'outer' curve is also less than that of the 'inner' curve. Therefore, if A' and A_1' are the return points on the positive x_2-axis of the trajectories through A and A_1, then

$$\Delta(A_1, A_1') < \Delta(A, A').$$

It follows that f is a decreasing function, i.e. if $0 < a < a_1$ then $f(a_1) < f(a)$. The function f also has the properties:

1. The origin of system (5.5) is an unstable focus (by the linearization theorem) and so $f(a)$ is *positive* for a small when the point A is close to the origin.
2. For large a, $f(a)$ is *negative*. As a increases the trajectory through A to D is displaced to the right in Fig. 5.2 because $a = OA$. Both the interval of integration and the integrand of (5.12) therefore, strictly increase with a. It follows that $\Delta(B, C)$, and hence $\Delta(A, A')$, tend to $-\infty$ as $a \to \infty$.
3. The function f is continuous because solution curves depend continuously on their initial points over finite intervals of the variable t.

Therefore, there exists $a_0 \in \mathrm{R}$ such that $f(a_0) = 0$ and so there is a closed trajectory through the point A_0 on the positive x_2-axis where $OA_0 = a_0$. Furthermore, the trajectories on either side approach the closed trajectory as $t \to \infty$ since $f(a) \gtrless 0$ for $a \lessgtr a_0$. □

We can now examine the generalized Van der Pol equation (5.3). The hypotheses in Theorem 5.1.2 on the functions $G(x)$, $h(x)$ generalize the important aspects of the corresponding coefficient functions in (5.4). These enable the existence of an attracting limit cycle to be proved by the same methods as in Theorem 5.1.1.

Theorem 5.1.2. The phase portrait of the equation

$$\ddot{x} + g(x)\dot{x} + h(x) = 0 \tag{5.17}$$

where:

(a) $G(x) = \int_0^x g(u)\mathrm{d}u$ is an even function which is zero only at $x = 0$ and $x = \pm\mu$, $\mu > 0$;

(b) $G(x) \to \infty$ as $x \to \infty$ monotonically for $x > \mu$;

(c) $h(x)$ is an odd function and $h(x) > 0$ for $x > 0$;

consists of a unique attracting limit cycle surrounding an unstable focus at the origin.

With these conditions satisfied Equation (5.3) is known as a *Liénard equation.* This result provides sufficient but not necessary conditions for the limit cycle behaviour to occur.

5.2 Regularization and some economic models

The idea that constrained dynamical equations can sometimes be regularized (see Section 4.4) has not escaped the notice of mathematical model-makers. In this section we describe an interesting sequence of models of business cycles constructed by Goodwin (1951). A crude model, involving constraints and jump assumptions, is first developed in order to capture the essentially non-linear nature of the problem. Mechanisms leading to regularization of the dynamical equations are then introduced as refinements; culminating in a Rayleigh-like equation (see Exercise 4.28) in the final model. This process of refinement is instructive from the stand point of modelling and it is particularly interesting to see how the 'folded characteristic' is engineered.

We are to be concerned with models of an economy with a view to accounting for fluctuations corresponding to the 'booms' and 'depressions' that have occurred in many national economies. Let us begin by introducing the variables involved and some basic relationships between them.

At any time t, the economy is taken to have 'capital stock', K consisting of plant, machinery, etc. which is changing at a rate $\dot{K}$ equal to the net investment/deterioration at that time. The economy has an 'income' from its 'output' Y and a 'consumption' C. These quantities are taken to be related by

$$C = \alpha Y + \beta \tag{5.18}$$

$$Y = C + \dot{K}, \tag{5.19}$$

where α and β are real constants such that $\alpha < 1$ and $\beta < C$. Equation (5.18) states that the consumption of the economy depends on its output in a linear way. Equation (5.19) implies that output is either consumed or invested. Further, the capital stock, K, is assumed to be controlled by a desire to maintain it at a level proportional to the output. Thus if R is the capital stock requirement at time t then

$$R = \gamma Y \tag{5.20}$$

where $\gamma \in \mathsf{R}$.

(*a*) *Model* 1

It is now possible to suggest a way in which cycles in the economy can occur. The basic equations (5.18) and (5.19) imply

$$Y = \frac{\beta + \dot{K}}{1 - \alpha}. \tag{5.21}$$

It follows that periodic behaviour in Y (and K) can be generated by oscillations in investment, $\dot{K}$. These oscillations, in turn, arise from the desire to make K equal to R, the desired level of capital stock.

Let us assume a rather extreme investment policy and suppose that

$$\dot{K} = \begin{cases} \kappa_1 > 0, & K < R \\ 0, & K = R \\ -\kappa_2 < 0, & K > R \end{cases} \tag{5.22}$$

where κ_1, κ_2 are independent of t. This corresponds to maximum investment if capital stock is less than that desired and zero investment (i.e. stock deteriorates at rate κ_2) if the desired level is exceeded. Normally, we would expect the rate at which plant can be created by maximum investment to be greater than its rate of removal by deterioration and obsolescence, i.e.

$$\kappa_1 > \kappa_2. \tag{5.23}$$

Equations (5.20)–(5.22) imply

$$R = \begin{cases} R_1 = \gamma(\beta+\kappa_1)/(1-\alpha), & K < R \\ R_0 = \gamma\beta/(1-\alpha), & K = R \\ R_2 = \gamma(\beta-\kappa_2)/(1-\alpha), & K > R. \end{cases} \tag{5.24}$$

Let $R_2 < K < R_1$ so that $R = R_1$ at $t = 0$. Investment is therefore equal to $\kappa_1 > 0$, K increases and Y remains constant, as shown in Fig. 5.4, until $K = R_1$. Here R changes to R_0 since $K = R$; now $K = R_1 > R = R_0$ so R switches straight through to R_2. Thus, effectively R changes instantaneously from R_1 to R_2 and $\dot{K}$ changes from κ_1 to $-\kappa_2$. The output Y falls discontinuously at this same instant by (5.21). Now K decreases until $K = R_2$. A similar argument shows that R switches to R_1 so that $K = R_2 < R = R_1$ and $\dot{K}$ becomes $\kappa_1 > 0$ again. Capital stock, K, again rises to R_1 and the cycle is complete. Thus K and Y undergo oscillations as shown in Fig. 5.4.

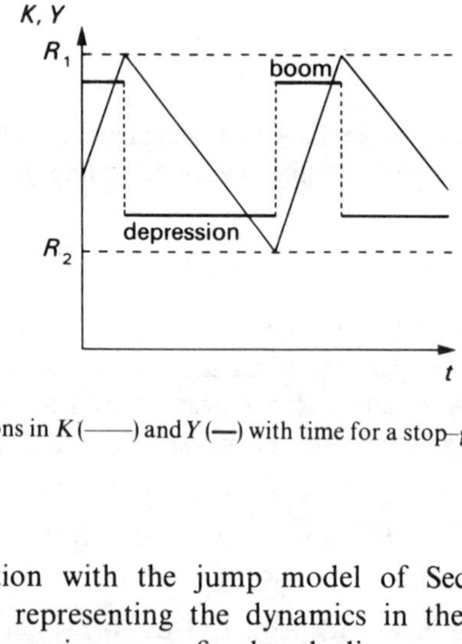

Fig. 5.4. Osillations in K(——) and Y(—) with time for a stop–go investment policy.

The connection with the jump model of Section 4.4 can be emphasized by representing the dynamics in the $K\dot{K}$-plane (see Fig. 5.5). The dynamics are confined to the line segments BC and DA, where, respectively, $\dot{K} = \kappa_1$ and $\dot{K} = -\kappa_2$. The jumps from A to B and C to D correspond to the discontinuities in Y shown in Fig. 5.4.

As a mathematical description of a business cycle, this first model has some success. During periods of investment, output is high and the economy booms. When investment is zero output is low and the economy is in a depression. However, the model also has many

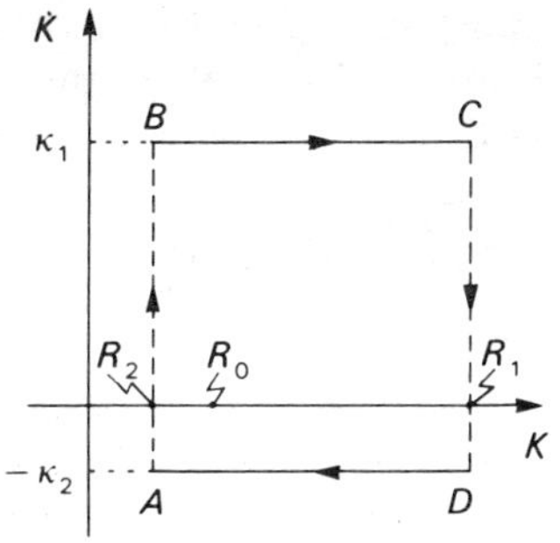

Fig. 5.5. Phase plane representation of the state of the economy. Jumps are represented by dashed lines.

weaknesses. For instance, discontinuous changes in investment and immediate response by output to investment (implied by (5.21)) are unrealistic. Further, $\kappa_1 > \kappa_2$ implies that the depressions are long compared with the booms; this is not what is observed. There is no growth of the economy since output, capital stock, etc. all repeatedly return to previously held values.

Space does not permit us to deal with all of these points in detail; here our aim is to develop the connection with our earlier discussion. We will therefore look at only two items:

1. the response of output to investment;
2. discontinuous changes in investment.

(*b*) *Model* 2

In dealing with item 1 it is apparent that we must modify (5.21) in such a way that discontinuous changes in Y do *not* result, even if discontinuous changes in $\dot{K}$ should occur. This can be achieved by replacing (5.21) by

$$Y = \frac{1}{1-\alpha}(\beta + \dot{K} - \varepsilon\dot{Y}) \tag{5.25}$$

where ε is a positive constant.

It is easily shown that the new term introduces a lag in the output response. Observe that (5.25) implies

$$\varepsilon\dot{Y} + (1-\alpha)Y = \beta + \kappa_1, \qquad K < R \tag{5.26}$$

$$\varepsilon\dot{Y} + (1-\alpha)Y = \beta - \kappa_2, \qquad K > R. \tag{5.27}$$

Now suppose a depression ends at $t = t_1$ and that there is an instantaneous switch from (5.27) to (5.26) then the solution of the latter equation shows that during the subsequent boom

$$Y(t) = \frac{\beta + \kappa_1}{1 - \alpha}\left\{1 - \exp\left[\frac{\alpha - 1}{\varepsilon}(t - t_1)\right]\right\} + Y(t_1)\exp\left[\frac{\alpha - 1}{\varepsilon}(t - t_1)\right]. \tag{5.28}$$

It follows that Y does *not* increase instantaneously to $(\beta + \kappa_1)/(1 - \alpha)$; rather it approaches this value as $t \to \infty$. Of course, from the practical point of view, the time taken for $Y(t)$ to approach this limit to any given tolerance is completely controlled by the parameter ε. Similarly (5.27) 'smooths' the discontinuous fall of Y (see Fig. 5.4) at the end of a boom.

Having assured ourselves that the output response lags behind changes in investment, we must turn our attention to item (2) and remove the discontinuous changes in investment. We must aim to soften the sudden change from $\dot{K} = \kappa_1$ to $\dot{K} = -\kappa_2$ (and vice versa) when K becomes equal to R.

To do this, we will consider the part of the investment that is *induced* by a change in output. This comes from our desire to keep the capital stock $K = R = \gamma Y$, the capital stock requirement. A change in Y produces a change in R and this requires a change in K (i.e. $\dot{K} \neq 0$) if $K = R$ is to be maintained. Clearly if we were completely successful in our desires we should have $\dot{K} = \gamma\dot{Y}$. Unfortunately, we know this cannot be true for all $\dot{Y} \in \mathrm{R}$ because $\dot{K}$ has an upper limit of κ_1 and a lower limit of $-\kappa_2$.

We, therefore, suggest that $\dot{K} = \psi(\dot{Y})$ where $\psi(\dot{Y})$ is shown in Fig. 5.6. For sufficiently small $\dot{Y}$ investment is able to keep pace with requirements and $K = \gamma Y$ is maintained. Finally investment reaches its extreme value and capital stock fails to satisfy requirements.

This argument means that $\dot{K}$ should be written

$$\dot{K} = L + \psi(\dot{Y}) \tag{5.29}$$

where $\psi(\dot{Y})$ is the induced investment and L represents other non-induced investment contributions. Thus (5.25) should be replaced by

$$Y = \frac{1}{1 - \alpha}[\beta + L + \psi(\dot{Y}) - \varepsilon\dot{Y}]. \tag{5.30}$$

The function $\psi(\dot{Y}) - \varepsilon\dot{Y}$ can be seen in Fig. 5.6 when $\varepsilon < \gamma$; its 'folded' or 'cubic-like' shape is clearly apparent.

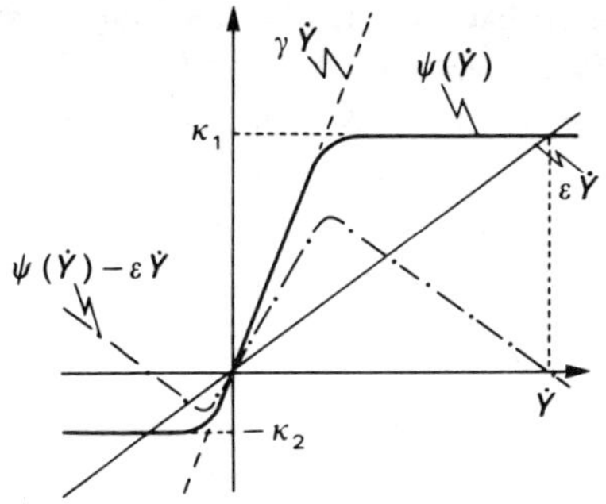

Fig. 5.6. Induced investment $\psi(\dot{Y})$ is approximated by the ideal $\gamma\dot{Y}$ for $\dot{Y}$ near zero but is ultimately limited by κ_1 and $-\kappa_2$. The functions $\varepsilon\dot{Y}$ and $\psi(\dot{Y})-\varepsilon\dot{Y}$ are also shown. Observe the folded nature of the latter.

To obtain a plot of Y against $\dot{Y}$ we must shift $\psi(\dot{Y})-\varepsilon\dot{Y}$, in Fig. 5.6, upwards by $\beta+L$ and scale by $(1-\alpha)$. Provided $\beta+L$ is sufficiently large we obtain a plot of the form shown in Fig. 5.7. This curve, along with a jump assumption completely determines the behaviour of this second model.

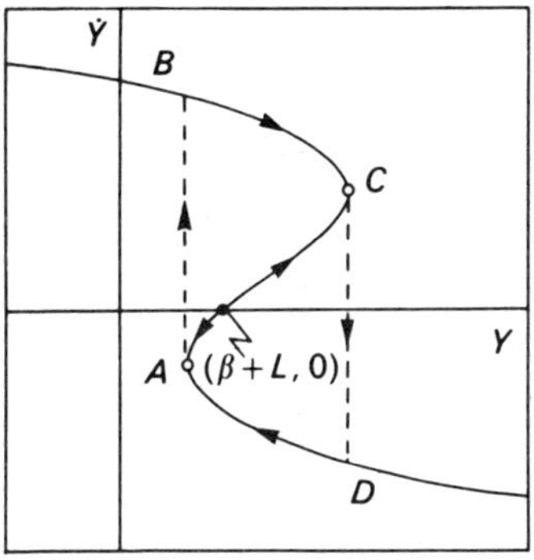

Fig. 5.7. Phase plane representation of Model 2. The dynamics are confined to the folded characteristic.

All states of the model lie on the curve and the sign of $\dot{Y}$ determines where Y is increasing or decreasing. Thus the state of the system must move in the direction of the arrows. The point $(\beta+L, 0)$ is consequently a repelling fixed point of the system. Points C and A are candidates for a jump assumption by analogy with Fig. 4.30. Assuming jumps A to B and C to D, we obtain relaxation oscillations in Y.

(c) *Model* 3

The jump assumption can be removed by moving to a two-dimensional system with a limit cycle. This can be obtained by recognizing that there is a lag between the decision to make an investment and the actual appearance of the investment outlays. This means that the induced investment at time t is actually determined by $\dot{Y}(t-\theta)$, where θ is the implementation lag, rather than $\dot{Y}(t)$.

We, therefore, replace (5.30) by

$$\varepsilon\dot{Y}(t)+(1-\alpha)Y(t)-\psi(\dot{Y}(t-\theta))=\beta+L \tag{5.31}$$

or putting $\tau=t-\theta$,

$$\varepsilon\dot{Y}(\tau+\theta)+(1-\alpha)Y(\tau+\theta)-\psi(\dot{Y}(\tau))=\beta+L. \tag{5.32}$$

Expanding in powers of θ and retaining only first-order terms we find

$$\varepsilon\theta\ddot{Y}(\tau)+[\varepsilon+(1-\alpha)\theta]\dot{Y}(\tau)-\psi(\dot{Y}(\tau))+(1-\alpha)Y(\tau)=\beta+L. \tag{5.33}$$

If we assume $\beta+L$ is constant and take $y=Y-(\beta+L)/(1-\alpha)$, (5.33) can be written in the homogeneous form

$$\varepsilon\theta\ddot{y}+[\varepsilon+(1-\alpha)\theta]\dot{y}-\psi(\dot{y})+(1-\alpha)y=0.$$

Finally, Goodwin takes $x=\sqrt{[(1-\alpha)/\varepsilon\theta]}\,y$, $t=\sqrt{[(1-\alpha)/\varepsilon\theta]}\,\tau$ and

$$\chi(\dot{x})=\{[\varepsilon+(1-\alpha)\theta]\dot{x}-\psi(\dot{x})\}/\sqrt{[(1-\alpha)\varepsilon\theta]}$$

to obtain the form

$$\ddot{x}+\chi(\dot{x})+x=0. \tag{5.34}$$

When $[\varepsilon+(1-\alpha)\theta]<\gamma$, $\chi(\dot{x})$ is a 'cubic-like' function, (5.34) is a Rayleigh-like equation and a stable limit cycle occurs. Figure 5.8 shows the phase portrait of (5.34) for typical values of the parameters.

5.3 The Zeeman models of heartbeat and nerve impulse

These models (Zeeman, 1973) are examples of the geometrical approach to modelling with differential equations. They are

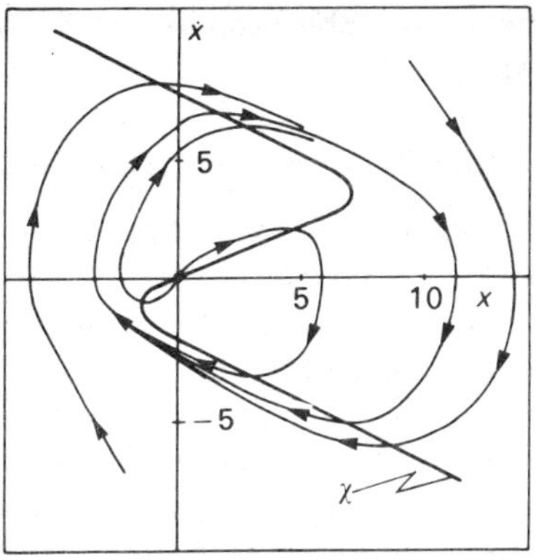

Fig. 5.8. Phase portrait for (5.34) when $\varepsilon = 0.5$, $\theta = 1.0$, $\alpha = 0.6$, $\gamma = 2.0$, $\kappa_1 = 9.0 \times 10^9$, $\kappa_2 = 3.0 \times 10^9$. Details of the origin of these numerical values are given in Goodwin (1951).

constructed from a purely qualitative description of the dynamics of the biological mechanisms. The particular differential equations chosen are merely the 'simplest' ones having the required dynamics. There are no specific conditions imposed on their form by any mechanisms whereby the dynamics arise. Only a variety of qualitative features of phase portraits are invoked in order to produce a mathematical description of the heartbeat and nerve impulse. This is in direct contrast to the modelling procedure in Section 5.6 where assumptions about the mechanisms involved completely determine the form of the dynamical equations.

The heart is predominantly in one of two states; relaxed (diastole) or contracted (systole). Briefly, in response to an electrochemical trigger each muscle fibre contracts rapidly, remains contracted for a short period and then relaxes *rapidly* to its stable relaxed state and so on. In contrast, nerve impulses have different dynamic behaviour. The part of the nerve cell which transmits messages is the axon. The operative quantity here is the electrochemically stimulated potential between the inside and outside of the axon. In the absence of stimuli the axon potential remains at a constant rest potential. When a message is being transmitted the axon potential changes sharply and then returns *slowly* to its rest potential.

These actions have three qualitative features in common and they form the basis of the models. They are:

(a) the existence of a stable equilibrium, to which the system returns periodically;
(b) a mechanism for triggering the action; and
(c) a return to equilibrium after the action is completed.

In this context the main difference between the heartbeat and nerve impulse results from the way in which item (c) takes place.

To construct a model of heartbeat with the above qualities we need to be clear about their interpretation in terms of phase portrait behaviour. Property (a) is interpreted as a stable fixed point in the phase portrait. Property (b) assumes there is a device for periodically moving the state of the system from the fixed point to some other state point. The trajectory through this nearby 'threshold' state then executes the fast action followed by the fast return to equilibrium as required in (c). To show how this can be done we discuss a sequence of models of increasing complexity.

We will use Zeeman's (1973) notation and consider

$$\dot{x} = -\lambda x, \qquad \dot{b} = -b \tag{5.35}$$

where λ is much larger than 1 (see Fig. 5.9). All trajectories other than those on the b-axis are almost parallel to the principal direction of $-\lambda$, the fast eigenvalue, and eventually move in a direction closely parallel to the principal direction of the slow eigenvalue -1. The equations $\dot{x} = -\lambda x$, $\dot{b} = -b$ are called the fast and slow equations

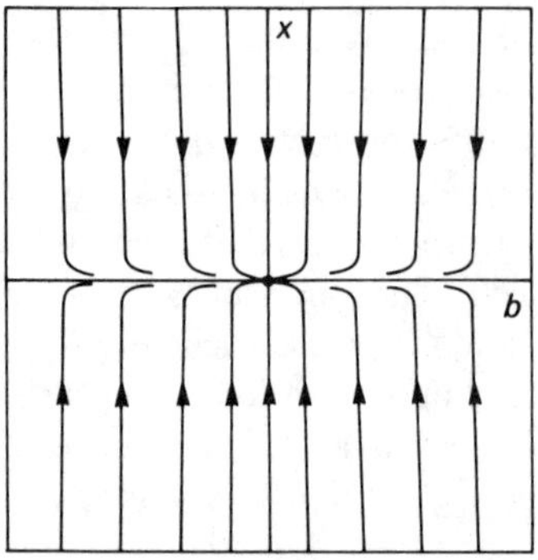

Fig. 5.9. The stable node of the system (5.35) with fast $(-\lambda)$ and slow (-1) eigenvalues.

respectively. Another example of this type of behaviour is

$$\varepsilon\dot{x} = x - b, \qquad \dot{b} = x \tag{5.36}$$

where ε is positive and much smaller than 1. The eigenvalues are approximately $1/\varepsilon$ and 1 and the principal directions are approximately along $b = 0$ and $b = (1-\varepsilon)x$. The fast equation is $\varepsilon\dot{x} = x - b$ as this contains the small term ε which gives the fast eigenvalue $1/\varepsilon$. The phase portrait of system (5.36) is sketched in Fig. 5.10. As in the previous example the fast and slow movements are almost parallel to the principal directions of the node.

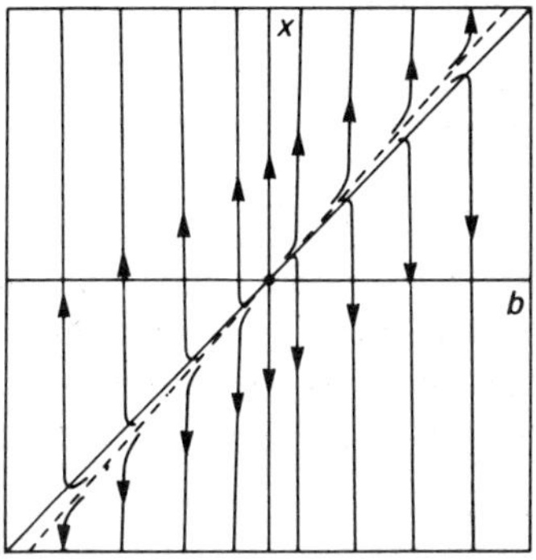

Fig. 5.10. The stable node $\varepsilon\dot{x} = x - b$, $\dot{b} = x$ with fast movement parallel to the x-axis and slow movement close to $b = (1-\varepsilon)x$.

Now consider the system

$$\varepsilon\dot{x} = x - x^3 - b, \qquad \dot{b} = x \tag{5.37}$$

for ε positive and small compared to 1. Equation (5.37) can be regarded as a modification of (5.36) with the linear term x replaced by the cubic function $x - x^3$. In fact the linearization of (5.37) at the fixed point (0, 0) is just (5.36). The $\dot{x} = 0$ isocline for the non-linear system (5.37) is now a cubic curve $b = x - x^3$ as illustrated in Fig. 5.11. The fast movement is almost parallel to the x-direction and the slow movement occurs close to the characteristic curve $b = x - x^3$. System (5.37) is of Liénard type as discussed in Sections 4.4 and 5.1. The shape of the limit cycle and some neighbouring trajectories are sketched in Fig. 5.11.

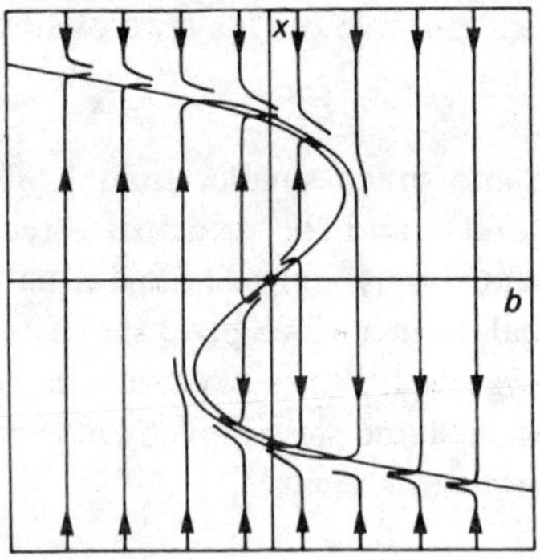

Fig. 5.11. The phase portrait of the Lienard system $\varepsilon\dot{x} = x - x^3 - b$, $\dot{b} = x$, where $0 < \varepsilon \ll 1$.

Zeeman (1973) makes the adjustment

$$\varepsilon\dot{x} = x - x^3 - b, \qquad \dot{b} = x - x_0 \tag{5.38}$$

to the Liénard system (5.37) to obtain a model of heartbeat. The value of x_0 is taken to be greater than $1/\sqrt{3}$, the value of x at A in Fig. 5.12. The system (5.38) then has a unique fixed point $E = (x_0, b_0)$, where $b_0 = x_0 - x_0^3$, on the upper fold of the characteristic $b = x - x^3$. The linearization of (5.38) at E is

$$\begin{bmatrix} \dot{x} \\ \dot{b} \end{bmatrix} = \begin{bmatrix} 1 - 3x_0^2 & -1 \\ 1 & 0 \end{bmatrix} \begin{bmatrix} x \\ b \end{bmatrix}$$

and therefore E is stable.

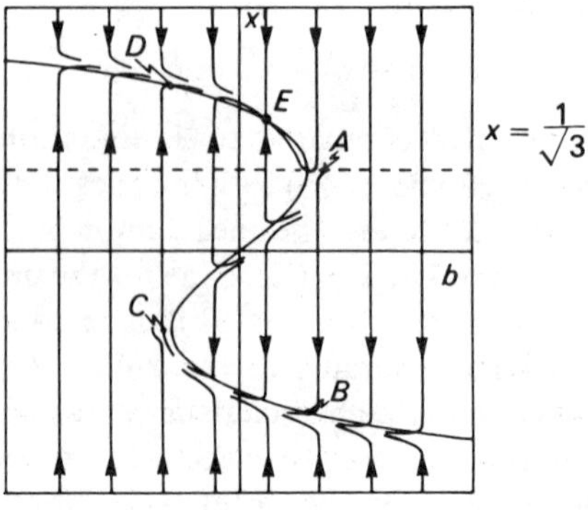

Fig. 5.12. The phase portrait of $\varepsilon\dot{x} = x - x^3 - b$, $\dot{b} = x - x_0$, where $0 < \varepsilon \ll 1$.

The variable x is interpreted as the length of a muscle fibre and b is an electrochemical control. The trigger mechanism moves the heart muscle from its equilibrium state E to the nearby state A (see Fig. 5.12). The muscle fibre then contracts rapidly as x decreases along the trajectory from A to B. A rapid relaxation of the muscle fibre occurs on the segment CD of the trajectory before returning to its relaxed state E. This behaviour is then repeated periodically by the action of the trigger to mimic the behaviour of the muscle fibres in a beating heart.

To obtain the slow return involved in a nerve impulse a model in R^3 is required. The need for a higher dimensional system of differential equations can be seen as follows. The fast action of a model like (5.38) arises from a folded characteristic curve. This essentially means that if there is a periodic orbit there must be a second fold and, consequently, a fast return. The introduction of a further dimension avoids this difficulty because the return does not have to take place in the vicinity of a fold. We are led, therefore, to models involving a characteristic *surface* in R^3.

The folded surface M around which the dynamics are constructed is given in (x, a, b)-space by

$$x^3 + ax + b = 0, \tag{5.39}$$

(see Fig. 5.13).

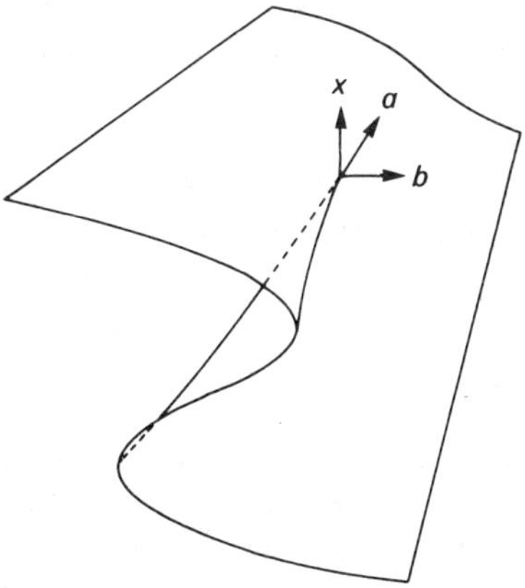

Fig. 5.13. The folded surface $x^3 + ax + b = 0$.

To see why Equation (5.39) gives the surface depicted here, it is instructive to consider various a = constant planes in R^3 and see how the surface M intersects these planes as a increases from negative values through zero to positive values. For a = constant, the equation $x^3 + ax + b = 0$ is a cubic curve in the x, b-coordinates. Figure 5.14 illustrates how the fold in the surface for $a < 0$ disappears for $a > 0$.

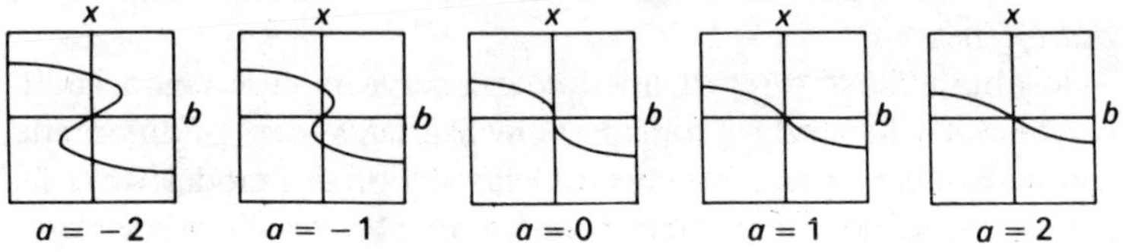

Fig. 5.14. Various a = constant sections of the surface M.

Differential equations which model fast action with slow return are given by

$$\begin{aligned} \varepsilon\dot{x} &= -(x^3 + ax + b) \\ \dot{a} &= -2x - 2a, \\ \dot{b} &= -a - 1 \end{aligned} \tag{5.40}$$

where ε is a small positive constant. As usual the fast equation gives large values of $\dot{x}$ for points not close to the surface M. This will give rise to fast movement in the x-direction in Fig. 5.16. For the slow return we shall see that the equations enable movement from the bottom sheet to the top sheet of M with no abrupt changes in any of the variables.

There is a unique fixed point E for system (5.40) at $x = 1$, $a = -1$, $b = 0$ on the surface M with linearization

$$\begin{bmatrix} \dot{x} \\ \dot{a} \\ \dot{b} \end{bmatrix} = \begin{bmatrix} -2/\varepsilon & -1/\varepsilon & -1/\varepsilon \\ -2 & -2 & 0 \\ 0 & -1 & 0 \end{bmatrix} \begin{bmatrix} x \\ a \\ b \end{bmatrix} \tag{5.41}$$

This linear system has eigenvalues $\frac{1}{2}(-1 \pm i\sqrt{3})$ and $-2/\varepsilon$, to the first order in ε, and so E is a stable fixed point. The fast eigenvalue gives rapid contraction towards E in the x-direction at points away from the surface M whilst the slow eigenvalues give spiralling towards E in the vicinity of M. A sketch of the phase portrait of (5.40) viewed from above M is given in Fig. 5.15, the line OF being the top fold of the surface M.

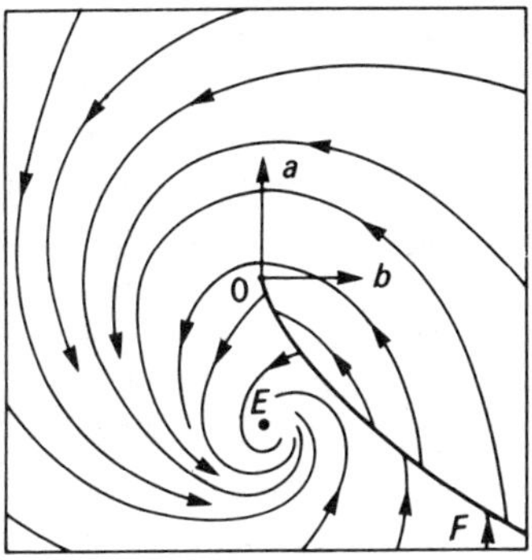

Fig. 5.15. A sketch of the phase portrait of (5.40) near to the surface M as viewed from a point on the positive x-axis.

With the help of Fig. 5.16, we can now describe the 'fast action–slow return' dynamic of system (5.40). We assume that the triggering process entails b increasing from zero at the equilibrium point E to some $b_c \geqslant 2/(3\sqrt{3})$ while x and a remain constant. The constraint on b_c ensures that the point $A = (1, -1, b_c)$ is to the right of the fold line OF (see Fig. 5.16). The dynamic given by (5.40) quickly

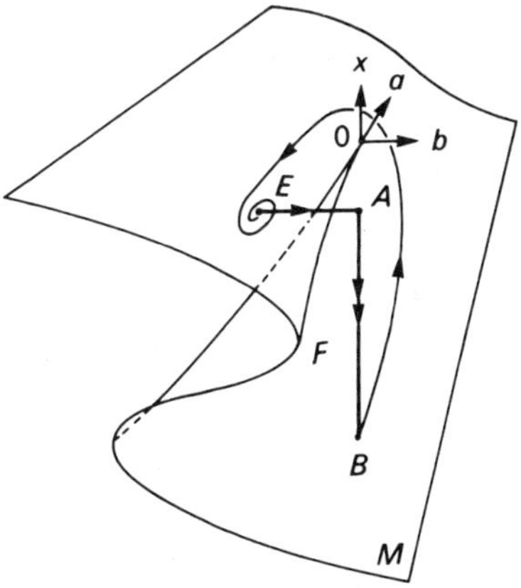

Fig. 5.16. The essential features of the dynamic on M.

changes x from 1 at A to the x-value at B which is almost vertically below A on the lower sheet of M. The trajectory through B then lies close to M and this ensures its relatively slow return to the top sheet before spiralling into E. The whole cycle can then be repeated.

To model the nerve impulse the variables x, a, b have to be suitably interpreted. During the transmission of a nerve impulse the ability of sodium ions to conduct through the cell undergoes a substantial and rapid *increase* and so sodium conductance is associated with $-x$ (recall x *decreases* rapidly). The trigger for this abrupt change was taken to be a small change in b. If b is taken to represent the potential across the cell membranes then a small increase in this potential is the nerve impulse triggering mechanism. Finally, the variable a represents potassium conductance. After the action, the conductance of potassium ions does appear to follow changes consistent with the trajectory through B. A slow rise in a is followed by a slow fall to equilibrium at $a = -1$, as the trajectory from A to E swings around O.

5.4 Liapunov functions

5.4.1 *Theory*

In Chapter 3 the main tool for discussing the behaviour of a non-linear system at one of its fixed points was the linearization theorem. The theorem gave conclusive results only when the fixed points were simple and not of centre type. However, in Section 3.5 it was shown how to categorize all fixed points into three types: asymptotically stable, neutrally stable and unstable. This section discusses a method for detecting these types of stability.

For example, suppose we wish to investigate the nature of the fixed point at the origin of the system

$$\dot{x}_1 = -x_1^3, \qquad \dot{x}_2 = -x_2^3. \tag{5.42}$$

The linearization theorem is of no use here as the linearized system is clearly non-simple. However, we can show that the origin is asymptotically stable by examining how the function $V(x_1, x_2) = x_1^2 + x_2^2$ changes along the trajectories of (5.42).

Let $\mathbf{x}(t) = (x_1(t), x_2(t))$ be any solution curve of system (5.42), then

$$\dot{V}(\mathbf{x}(t)) = \frac{\partial V}{\partial x_1}\dot{x}_1 + \frac{\partial V}{\partial x_2}\dot{x}_2 = -2(x_1^4 + x_2^4). \tag{5.43}$$

Therefore, $\dot{V}(\mathbf{x}(t))$ is negative at all points other than the origin of the x_1, x_2-plane and so $V(\mathbf{x}(t))$ decreases as t increases. This means that the phase point $\mathbf{x}(t)$ moves towards the origin with increasing t. In fact, $\dot{V}(\mathbf{x}(t)) < 0$ for $\mathbf{x}(t) \neq \mathbf{0}$ implies that $V(\mathbf{x}(t)) \to 0$ as $T \to \infty$ and hence $\mathbf{x}(t) \to \mathbf{0}$ as $t \to \infty$. Thus, the origin is an asymptotically stable fixed point of system (5.42).

The above example is a simple illustration of the use of a Liapunov function. To develop this idea we will need the following definitions.

Definition 5.4.1. A real-valued function $V: N \subseteq \mathsf{R}^2 \to \mathsf{R}$, where N is a neighbourhood of $\mathbf{0} \in \mathsf{R}^2$, is said to be *positive* (*negative*) *definite* in N if $V(\mathbf{x}) > 0$ ($V(\mathbf{x}) < 0$) for $\mathbf{x} \in N \backslash \{\mathbf{0}\}$ and $V(\mathbf{0}) = \mathbf{0}$.

Definition 5.4.2. A real-valued function $V: N \subseteq \mathsf{R}^2 \to \mathsf{R}$, where N is a neighbourhood of $\mathbf{0} \in \mathsf{R}^2$, is said to be *positive* (*negative*) *semi-definite* in N if $V(\mathbf{x}) \geqslant 0$ ($V(\mathbf{x}) \leqslant 0$) for $\mathbf{x} \in N \backslash \{\mathbf{0}\}$ and $V(\mathbf{0}) = 0$.

Definition 5.4.3. The derivative of $V: N \subseteq \mathsf{R}^2 \to \mathsf{R}$ along a parameterized curve given by $\mathbf{x}(t) = (x_1(t), x_2(t))$ is defined by

$$\dot{V}(\mathbf{x}(t)) = \frac{\partial V(\mathbf{x}(t))}{\partial x_1}\dot{x}_1(t) + \frac{\partial V(\mathbf{x}(t))}{\partial x_2}\dot{x}_2(t). \tag{544}$$

The function $V(x_1, x_2) = x_1^2 + x_2^2$ used in the introductory example is positive definite on R^2. This function is typical of the positive definite functions used in this section. Any continuously differentiable, positive definite function V has a continuum of closed level curves around the origin. Of course, such curves are not necessarily circular (see Fig. 5.17). However, provided $\dot{V}$ is negative on

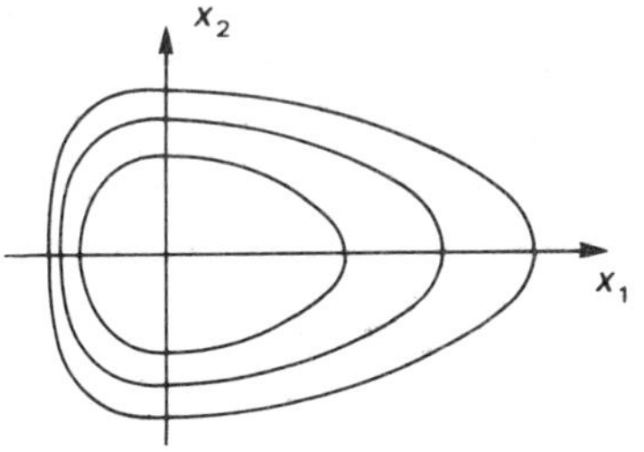

Fig. 5.17. The level curves $V(x_1, x_2) = C$ of the positive definite function $V(x_1, x_2) = x_1 - \log(1 + x_1) + x_2^2$ for $C = 0.5, 1.0, 1.5$.

a trajectory, then that trajectory must still move towards the origin, because V is decreasing. Observe that for any system $\dot{\mathbf{x}} = \mathbf{X}(\mathbf{x})$,

$$\dot{V}(\mathbf{x}(t)) = \left(\frac{\partial V}{\partial x_1}X_1 + \frac{\partial V}{\partial x_2}X_2\right) \tag{5.45}$$

is a function of x_1 and x_2 only and, for this reason, it is often denoted by $\dot{V}(\mathbf{x})$.

Theorem 5.4.1 (Liapunov stability theorem). Suppose the system $\dot{\mathbf{x}} = \mathbf{X}(\mathbf{x})$, $\mathbf{x} \in S \subseteq \mathsf{R}^2$ has a fixed point at the origin. If there exists a real-valued function V in a neighbourhood N of the origin such that:

(a) the partial derivatives $\partial V/\partial x_1, \partial V/\partial x_2$ exist and are continuous;

(b) V is positive definite;

(c) $\dot{V}$ is negative semi-definite;

then the origin is a *stable* fixed point of the system.

If (c) is replaced by the stronger condition

(c′) $\dot{V}$ is negative definite,

then the origin is an *asymptotically stable* fixed point.

Proof. Properties (a) and (b) imply that the level curves of V form a continuum of closed curves around the origin. Thus, there is a positive k such that $N_1 = \{\mathbf{x} \mid V(\mathbf{x}) < k\}$ is a neighbourhood of the origin contained in N. If $\mathbf{x}_0 \in N_1 \backslash \{\mathbf{0}\}$, then $\dot{V}(\phi_t(\mathbf{x}_0)) \leqslant 0$, for all $t \geqslant 0$, by (c) and $V(\phi_t(\mathbf{x}_0))$ is a decreasing function of t. Therefore, $V(\phi_t(\mathbf{x}_0)) < k$, for all $t \geqslant 0$, and N_1 is a positively invariant set. Consequently, by Definition 3.5.2 the fixed point is stable.

For case (c′) we obtain the asymptotic stability of the origin by the following argument. The function $V(\phi_t(\mathbf{x}_0))$ is decreasing and bounded below by zero so that it must tend to a limit as $t \to \infty$. It follows that $\lim_{t \to \infty} \dot{V}(\phi_t(\mathbf{x}_0)) = 0$ but, when $\dot{V}$ is negative definite, this can only occur if $\phi_t(\mathbf{x}_0) \to \mathbf{0}$ as $t \to \infty$. □

Definition 5.4.4. A function V satisfying hypotheses (a), (b) and (c) of Theorem 5.4.1 is called a *weak Liapunov function*. If (c) is replaced by (c′) then V is a *strong Liapunov function*.

Example 5.4.1. Prove that the function

$$V(y_1, y_2) = y_1^2 + y_1^2 y_2^2 + y_2^4, \qquad (y_1, y_2) \in \mathbb{R}^2 \tag{5.46}$$

is a strong Liapunov function for the system

$$\begin{aligned} \dot{x}_1 &= 1 - 3x_1 + 3x_1^2 + 2x_2^2 - x_1^3 - 2x_1 x_2^2 \\ \dot{x}_2 &= x_2 - 2x_1 x_2 + x_1^2 x_2 - x_2^3, \end{aligned} \tag{5.47}$$

at the fixed point (1, 0).

Solution. On introducing local coordinates y_1, y_2 at (1, 0), (5.47) becomes

$$\dot{y}_1 = -y_1^3 - 2y_1 y_2^2, \qquad \dot{y}_2 = y_1^2 \; y_2 - y_2^3. \tag{5.48}$$

The function V in (5.46) is positive definite and

$$\begin{aligned} \dot{V}(y_1, y_2) &= \frac{\partial V}{\partial y_1}\dot{y}_1 + \frac{\partial V}{\partial y_2}\dot{y}_2 \\ &= (2y_1 + 2y_1 y_2^2)(-y_1^3 - 2y_1 y_2^2) + \\ &\quad (2y_1^2 y_2 + 4y_2^3)(y_1^2 y_2 - y_2^3) \\ &= -2y_1^4 - 4y_1^2 y_2^2 - 2y_1^2 y_2^4 - 4y_2^6 \end{aligned}$$

is negative definite. Therefore, V, is a strong Liapunov function for (5.47). □

Example 5.4.2. Investigate the stability of the second-order equation

$$\ddot{x} + \dot{x}^3 + x = 0 \tag{5.49}$$

at the origin of its phase plane.

Solution. If $x_1 = x$ and $x_2 = \dot{x}$, then

$$\dot{x}_1 = x_2, \qquad \dot{x}_2 = -x_1 - x_2^3 \tag{5.50}$$

is the first-order system of (5.49). The derivative of the function $V(x_1, x_2) = x_1^2 + x_2^2$ along the trajectories of (5.50) is

$$\dot{V}(x_1, x_2) = -2x_2^4$$

and so $\dot{V}$ is only negative semi-definite. Hence by Theorem 5.4.1 the origin is a stable fixed point of system (5.50). □

In fact, asymptotic stability can be deduced for systems having a weak Liapunov function similar to that in Example 5.4.2. Observe that $\dot{V}(\mathbf{x})$ only fails to be negative away from the origin on the line $x_2 = 0$. On this line the components of the vector field given by (5.50) are $\dot{x}_1 = 0$, $\dot{x}_2 = -x_1$. Thus, all trajectories (except the origin) cross the line $x_2 = 0$ and $\dot{V}$ is only momentarily zero. At all other points in the plane it is negative. Under these circumstances the following theorem gives asymptotic stability.

Theorem 5.4.2. If there exists a weak Liapunov function V for the system $\dot{\mathbf{x}} = \mathbf{X}(\mathbf{x})$ in a neighbourhood of an isolated fixed point at the origin, then providing $\dot{V}(\mathbf{x})$ does not vanish identically on any trajectory, other than the fixed point itself, the origin is asymptotically stable.

Proof. Let V be defined on some neighbourhood N of the origin. Then by the same argument as in the proof of Theorem 5.4.1 there is a positively invariant neighbourhood $N_1 = \{\mathbf{x} \mid V(\mathbf{x}) < k\}$ of the origin contained in N. Theorem 3.9.1 states that the trajectory through any $\mathbf{x}_0 \in N_1$ either tends to a fixed point (at the origin) or spirals towards a closed orbit as $t \to \infty$. If the latter occurs then $\lim_{t \to \infty} \dot{V}(\phi_t(\mathbf{x}_0)) = 0$ and the continuity of $\dot{V}$ implies $\dot{V}(\mathbf{x}) = 0$ for all points $\mathbf{x}$ of the closed orbit. Thus $\dot{V}$ vanishes identically on the closed orbit contrary to hypothesis. We conclude that $\lim_{t \to \infty} \phi_t(\mathbf{x}_0) = \mathbf{0}$ and the origin is asymptotically stable. □

Example 5.4.3. Show that all trajectories of the system

$$\dot{x}_1 = x_2, \qquad \dot{x}_2 = -x_1 - (1 - x_1^2)x_2 \tag{5.51}$$

passing through points (x_1, x_2), with $x_1^2 + x_2^2 < 1$, tend to the origin with increasing t.

Solution. The function $V(x_1, x_2) = x_1^2 + x_2^2$ is a weak Liapunov function in the region $x_1^2 + x_2^2 < 1\,(\dot{V}(x_1, x_2) = -2x_2^2(1 - x_1^2))$. The function $\dot{V}$ vanishes only on the lines $x_2 = 0$ and $x_1 = \pm 1$. However, there are no trajectories of (5.51) which lie on these lines because on $x_2 = 0$, $\dot{x}_2 = -x_1 \not\equiv 0$ and on $x_1 = \pm 1$, $\dot{x}_1 = x_2 \not\equiv 0$. Therefore, by Theorem 5.4.2, the origin is asymptotically stable. Moreover the arguments used to prove Theorem 5.4.2 show that any trajectory $\phi_t(\mathbf{x}_0)$, $|\mathbf{x}_0| < 1$, has the property $\lim_{t \to \infty} \phi_t(\mathbf{x}_0) = \mathbf{0}$. □

The fact that the origin is an asymptotically stable fixed point of system (5.51) can be deduced by using the linearization theorem. However, the above solution provides an explicit 'domain of stability' $N = \{(x_1, x_2) | x_1^2 + x_2^2 < 1\}$. All trajectories through points of N approach the origin as t increases. The linearization theorem gives the existence of a domain of stability but no indication of its size.

Theorem 5.4.3. Suppose the system $\dot{\mathbf{x}} = \mathbf{X}(\mathbf{x})$ has a fixed point at the origin. If a real-valued, continuous function V exists such that:

(a) the domain of V contains $N = \{\mathbf{x} \,|\, |\mathbf{x}| \leqslant r\}$ for some $r > 0$;
(b) there are points arbitrarily close to the origin at which V is positive;
(c) $\dot{V}$ is positive definite; and
(d) $V(\mathbf{0}) = 0$,
then the origin is unstable.

Proof. We show that for every point $\mathbf{x}_0$ in N, with $V(\mathbf{x}_0) > 0$, the trajectory $\phi_t(\mathbf{x}_0)$ leaves N for sufficiently large positive t. By hypothesis, such points can be chosen arbitrarily close to the origin and therefore the origin is unstable.

Given r_1, such that $0 < r_1 < r$, there is a point $\mathbf{x}_0 \neq 0$, with $|\mathbf{x}_0| < r_1$ and $V(\mathbf{x}_0) > 0$. The function $\dot{V}$ is positive definite in N and so $V(\phi_t(\mathbf{x}_0))$ is an increasing function of t. Therefore, the trajectory $\phi_t(\mathbf{x}_0)$ does not approach the origin as t increases. Hence $\dot{V}(\phi_t(\mathbf{x}_0))$ will be bounded away from zero, i.e. there exists a positive K such that $\dot{V}(\phi_t(\mathbf{x}_0)) \geqslant K$ for all positive t. If we assume the trajectory $\phi_t(\mathbf{x}_0)$ remains in N, then

$$V(\phi_t(\mathbf{x}_0)) - V(\mathbf{x}_0) \geqslant Kt, \tag{5.52}$$

for all positive t and $V(\phi_t(\mathbf{x}_0))$ becomes arbitrarily large in N. This contradicts the hypothesis that V is a continuous function defined on the *closed* and *bounded* set N. Thus the trajectory $\phi_t(\mathbf{x}_0)$ must leave N as t increases. □

Example 5.4.4. Show that the system

$$\dot{x}_1 = x_1^2, \qquad \dot{x}_2 = 2x_2^2 - x_1 x_2 \tag{5.53}$$

is unstable at the origin by using the function

$$V(x_1, x_2) = \alpha x_1^3 + \beta x_1^2 x_2 + \gamma x_1 x_2^2 + \delta x_2^3, \tag{5.54}$$

for a suitable choice of constants $\alpha, \beta, \gamma, \delta$.

Solution. The derivative of V along the trajectories of system (5.53) is

$$\begin{aligned}\dot{V}(x_1, x_2) = {} & 3\alpha x_1^4 + \beta x_1^3 x_2 \\ & + (2\beta - \gamma) x_1^2 x_2^2 + (4\gamma - 3\delta) x_1 x_2^3 \\ & + 6\delta x_2^4.\end{aligned} \tag{5.55}$$

Observe that if we choose $\alpha = \frac{1}{3}$, $\beta = 4$, $\gamma = 2$, $\delta = \frac{4}{3}$ then the various terms of $\dot{V}$ can be grouped together to form

$$\begin{aligned}\dot{V}(x_1, x_2) &= x_1^4 + 4x_1^3 x_2 + 6x_1^2 x_2^2 + 4x_1 x_2^3 + 8x_2^4 \\ &= (x_1 + x_2)^4 + 7x_2^4\end{aligned} \tag{5.56}$$

which is clearly positive definite. The function V is given by

$$V(x_1, x_2) = \tfrac{1}{3}x_1^3 + 4x_1^2 x_2 + 2x_1 x_2^2 + \tfrac{4}{3}x_2^3. \tag{5.57}$$

This function has the property that $V(x_1, x_2) = \frac{1}{3}x_1^3$ when $x_2 = 0$, and so points arbitrarily close to the origin on the x_1-axis can be found for which V is positive. It follows that the origin is an unstable fixed point by Theorem 5.4.3. □

5.4.2 *A model of animal conflict*

Suppose we wish to model the ritualized conflicts which occur within a species when, for example, there is competition for mates, territory or dominance. Conflict occurs when two individuals confront one another and we will suppose that it consists of three possible elements:

(a) display;
(b) escalation of a fight; or
(c) running away.

The population to be modelled is taken to consist of individuals who respond to confrontation in one of a finite number of ways. Suppose that each individual adopts one of the strategies given in the following table.

An individual playing strategy i against an opponent playing j receives a 'pay-off' a_{ij}. This pay-off is taken to be related to the individuals capability to reproduce (i.e. the greater the pay-off, the

Index i	Strategy	Initial tactic	Tactic if opponent escalates
1	Hawk (H)	escalate	escalate
2	Dove (D)	display	run away
3	Bully (B)	escalate	run away

greater the number of off-spring). Assuming that only pure strategies are played (i.e. an individual is always true to type and always plays the same strategy) and that individuals breed true (i.e. off-spring play the same strategies as their parent), the model is able to determine the evolution of the three sections of the population.

Let x_i be the proportion of the population playing strategy i. It follows that

$$\sum_{i=1}^{3} x_i = 1 \tag{5.58}$$

and

$$x_i \geqslant 0, \qquad i = 1, 2, 3. \tag{5.59}$$

The pay-off to an individual playing i against the rest of the population is

$$\sum_j a_{ij} x_j = (\boldsymbol{A}\boldsymbol{x})_i, \tag{5.60}$$

where $\boldsymbol{A}$ is the 'pay-off matrix'. The average pay-off to an individual is

$$\sum_i x_i (\boldsymbol{A}\boldsymbol{x})_i = \boldsymbol{x}^T \boldsymbol{A}\boldsymbol{x}. \tag{5.61}$$

The 'advantage' of playing i is therefore

$$(\boldsymbol{A}\boldsymbol{x})_i - \boldsymbol{x}^T \boldsymbol{A}\boldsymbol{x}. \tag{5.62}$$

The per capita growth rate of the section of the population playing i is taken to be proportional to this advantage. A suitable choice of units of time then gives

$$\dot{x}_i = x_i((\boldsymbol{A}\boldsymbol{x})_i - \boldsymbol{x}^T \boldsymbol{A}\boldsymbol{x}), \tag{5.63}$$

$i = 1, 2, 3$. These equations only represent the state of the population for points in R^3 which satisfy (5.58) and (5.59), i.e. on the region, Δ, in Fig. 5.18.

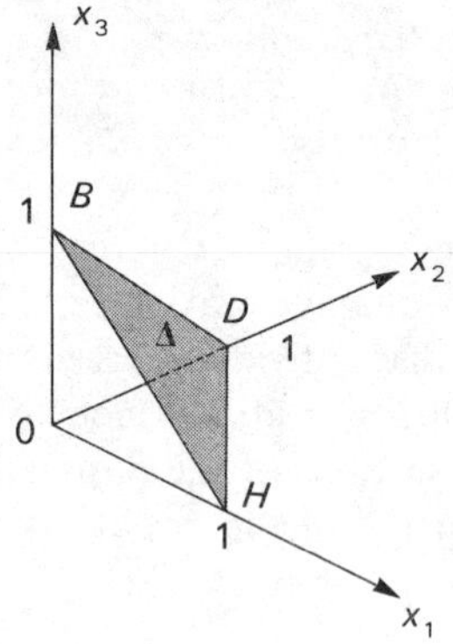

Fig. 5.18. The dynamics of the animal conflict model (5.63) are confined to $\Delta = \{(x_1, x_2, x_3) | \sum_{i=1}^{3} x_i = 1, \ x_i \geqslant 0, \ i = 1, 2, 3\}$. This is the plane triangular 'constraint surface' *BDH*.

We can obtain a pay-off matrix by awarding scores, at each confrontation, according to: win = 6, lose = 0, injury = -10, time wasted = -1. The actual values chosen here are not important; it is their signs and the order of their magnitudes that are significant. A hawk meeting a dove or a bully simply wins so $a_{12} = a_{13} = 6$. If two hawks meet they fight until one is injured. Each hawk has an equal probability of winning and so the expected gain is $\frac{1}{2}(6-10) = -2 = a_{11}$. A dove meeting either a hawk or a bully loses so that $a_{21} = a_{23} = 0$, but two doves continue to display for some time before one gives up. Thus $a_{22} = \frac{1}{2}(6+0) - 1 = 2$. Finally, bullies lose against hawks ($a_{31} = 0$), win against doves ($a_{32} = 6$) and have a 50% chance of winning against bullies $[a_{33} = \frac{1}{2}(6+0) = 3]$. Thus

$$A = \begin{bmatrix} -2 & 6 & 6 \\ 0 & 2 & 0 \\ 0 & 6 & 3 \end{bmatrix}. \tag{5.64}$$

It is also useful to note that the advantage of a strategy is unchanged by the addition of a constant to a column of A. Thus A can be simplified by reducing its diagonal elements to zero by such column changes without changing the dynamical equation (5.63). Therefore we take

$$A = \begin{bmatrix} 0 & 4 & 3 \\ 2 & 0 & -3 \\ 2 & 4 & 0 \end{bmatrix}. \tag{5.65}$$

The dynamical equations (5.63) are, in fact, only part of a more substantial model described by Zeeman (1979); his paper gives a clear account of the model and its dynamics. Our purpose here is merely to draw the readers' attention to models of this kind and, at the same time, to illustrate the value of the Liapunov approach when domains of stability are required.

Example 5.4.5. Show that the dynamical equations (5.63), with A given by (5.65), have a fixed point at $(x_1, x_2, x_3) = (\frac{3}{5}, 0, \frac{2}{5})$. Use the function

$$V(\mathbf{x}) = x_1^{3/5} x_3^{2/5}, \tag{5.66}$$

to show that this fixed point is asymptotically stable, with domain of stability

$$\mathring{\Delta} = \{(x_1, x_2, x_3) | x_1 + x_2 + x_3 = 1;\ x_1, x_2, x_3 > 0\}.$$

Find the remaining fixed points of (5.63), determine their nature and sketch the phase portrait on Δ. What happens to a population consisting entirely of hawks and doves if a mutant bully appears?

Solution. To check that $\mathbf{x} = (\frac{3}{5}, 0, \frac{2}{5})$ is a fixed point of (5.63) we note that $\boldsymbol{x}^T A \boldsymbol{x} = 5(\frac{3}{5})(\frac{2}{5}) = \frac{6}{5}$. For $i = 1$ and 3 $(A\boldsymbol{x})_i = \frac{6}{5}$ and hence $\dot{x}_1 = \dot{x}_3 = 0$. When $i = 2$, $\dot{x}_2 = 0$ since $x_2 = 0$.

We show that the point $(\frac{3}{5}, 0, \frac{2}{5})$ is asymptotically stable on $\mathring{\Delta}$ by using an argument of the Liapunov type. The level surfaces $V(x_1, x_2, x_3)$, cut the $x_2 = 0$ plane in hyperbolae and are invariant under translation parallel to the x_2-axis. A sketch is given in Fig. 5.19. The triangular surface Δ intersects these level surfaces in a system of curves illustrated in Fig. 5.20.

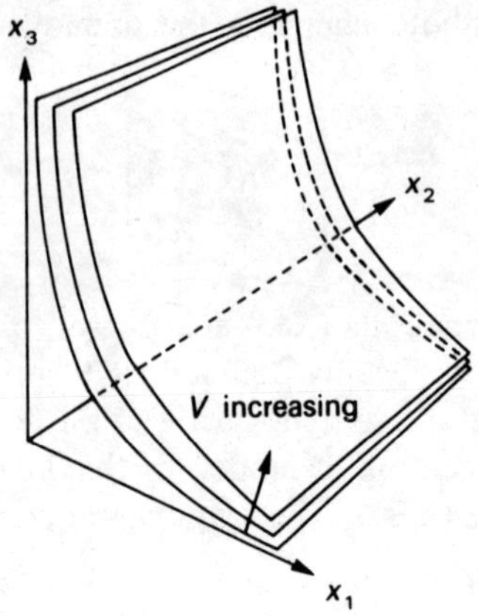

Fig. 5.19. The level surfaces of $V(x_1, x_2, x_3) = x_1^{3/5} x_3^{2/5}$ are generated by translating the hyperbolae $x_3 = Cx_1^{-3/2}$, C constant, in the x_2-direction.

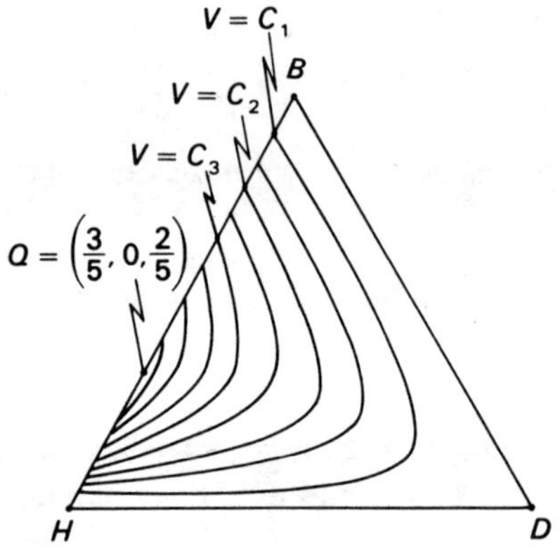

Fig. 5.20. The level curves of $V = x_1^{3/5} x_3^{2/5}$ on $\mathring{\Delta}$ obtained by intersecting the level surfaces of V with $\mathring{\Delta}$. V takes its unique maximum value on Δ at Q so that $C_3 > C_2 > C_1$.

On $\mathring{\Delta}$ the derivative of V along the trajectories of (5.63) is

$$\begin{aligned}\dot{V}(\mathbf{x}) &= V(\mathbf{x})\left(\frac{3\dot{x}_1}{5x_1} + \frac{2\dot{x}_3}{5x_3}\right) \\ &= V(\mathbf{x})[(\tfrac{3}{5}, 0, \tfrac{2}{5})\boldsymbol{A}\boldsymbol{x} - \boldsymbol{x}^T\boldsymbol{A}\boldsymbol{x}] \\ &= V(\mathbf{x})[(1 - x_1 - x_3)(\tfrac{11}{5} - x_1 - x_3) \\ &\quad + 5(x_1 - \tfrac{3}{5})^2] \end{aligned} \tag{5.67}$$

Hence, $\dot{V}(\mathbf{x})$ is positive for $\mathbf{x} \in \mathring{\Delta}$ and V increases along trajectories of (5.63) as t increases.

Using a similar argument to that given in Theorem 5.4.1, we can conclude that along any trajectory in $\mathring{\Delta}$ the function V *increases* to its *maximum* at Q (whereas V decreases to a minimum in the proof of Theorem 5.4.1). Thus all trajectories in $\mathring{\Delta}$ approach Q as t increases. It also follows that $\mathring{\Delta}$ is a subset of the domain of stability of Q and therefore cannot contain any fixed points. *All* the fixed points of (5.63) therefore appear on the boundary of Δ. On HB ($x_2 = 0$), the equations $\dot{x}_1 = 0$, $\dot{x}_3 = 0$ are given by $x_1(3x_3 - 5x_1x_3) = 0$ and $x_3(2x_1 - 5x_1x_3) = 0$, respectively. Thus, besides Q, there are fixed points at $H = (1, 0, 0)$ and $B = (0, 0, 1)$. Similarly, on BD ($x_1 = 0$) and HD ($x_3 = 0$) we find that $D = (0, 1, 0)$ and $P = (\frac{2}{3}, \frac{1}{3}, 0)$ are the only other fixed points on Δ.

We can obtain the behaviour of the trajectories on the boundary of Δ by observing that:

(a) on HB, $\dot{x}_1 > 0$ for $x_1 < \frac{3}{5}$ and $\dot{x}_1 < 0$ for $x_1 > \frac{3}{5}$;
(b) on BD, $\dot{x}_3 > 0$;
(c) on HD, $\dot{x}_2 > 0$ for $x_2 < \frac{1}{3}$ and $\dot{x}_2 < 0$ for $x_2 > \frac{1}{3}$.

Finally, we obtain the phase portrait shown in Fig. 5.21.

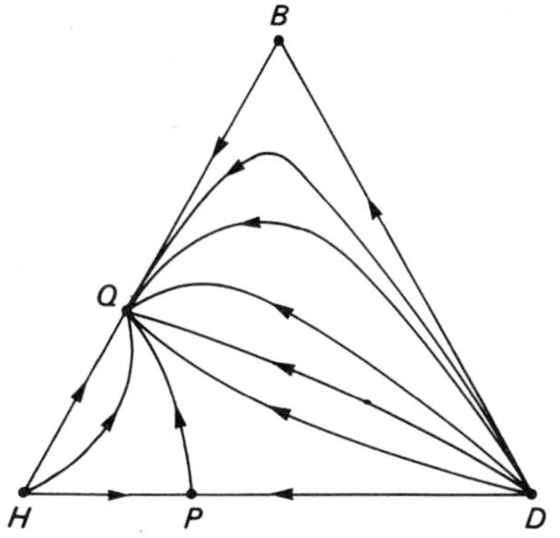

Fig. 5.21. The phase portrait on $\mathring{\Delta}$ of system (5.63) with A given by (5.65).

Suppose a mutant bully appears in a population consisting originally of only hawks and doves. The population mix is represented by a phase point in $\mathring{\Delta}$ near to HD. Since all trajectories in $\mathring{\Delta}$

approach Q as t increases, we conclude that the state of the population evolves to $x = (\frac{3}{5}, 0, \frac{2}{5})$ where the doves become extinct. □

5.5 Bifurcation of systems

5.5.1 *Some simple examples*

Dynamical equations frequently involve *parameters* or 'constants' as well as the dynamical variables. For example:

(a) the per capita growth rate, a, in the population equation

$$\dot{x} = ax; \tag{5.68}$$

(b) the natural frequency, ω_0, and damping constant k in the harmonic oscillator

$$\ddot{x} + 2k\dot{x} + \omega_0^2 x = 0;$$

(c) the quantity ε in the Van der Pol equation

$$\ddot{x} + \varepsilon\,(x^2 - 1)\dot{x} + x = 0;$$

are all 'parameters'.

The *bifurcation* of a differential equation is concerned with changes in the qualitative behaviour of its phase portrait as a parameter (or set of parameters) varies. For example, (5.68) has an attractor at $x = 0$ if $a < 0$ and a repellor if $a > 0$. As a increases through zero, the solutions change from decreasing to increasing functions of t. This differential equation is said to have a *bifurcation point* at $a = 0$. Similarly the system

$$\dot{x}_1 = \mu x_1, \qquad \dot{x}_2 = -x_2, \tag{5.69}$$

where $\mu \in \mathrm{R}$ undergoes a bifurcation at $\mu = 0$. Here qualitatively different phase portraits arise for $\mu < 0$, $\mu = 0$ and $\mu > 0$, as shown in Fig. 5.22.

For *any* $\mu < 0$ the phase portrait is a stable node; for $\mu = 0$ it is the phase portrait of a non-simple fixed point (cf. Fig. 2.6(b)); and for *any* $\mu > 0$, the phase portrait is a saddle.

Example 5.5.1. Find the different qualitative types of phase portrait obtained for the one-parameter system

$$\dot{x}_1 = \mu x_1 - x_2 - x_1(x_1^2 + x_2^2), \qquad \dot{x}_2 = x_1 + \mu x_2 - x_2(x_1^2 + x_2^2) \tag{5.70}$$

as μ increases from $-\infty$ to $+\infty$.

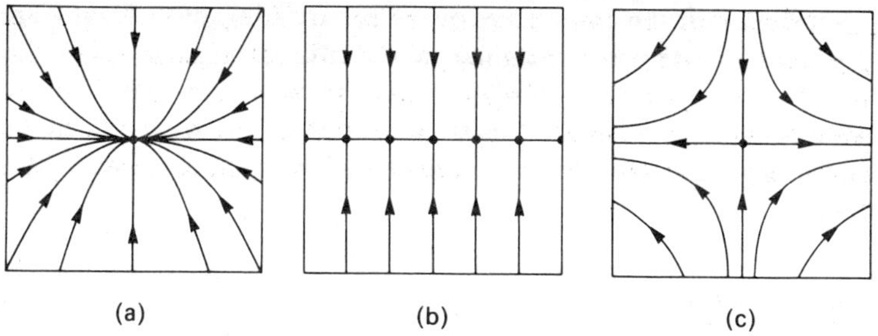

Fig. 5.22. Phase portraits of the parametrized system $\dot{x}_1 = \mu x_1$, $\dot{x}_2 = -x_2$: (a) $\mu < 0$; (b) $\mu = 0$; (c) $\mu > 0$.

Solution. The system (5.70) can be simplified by transforming to coordinates r, θ to obtain

$$\dot{r} = r(\mu - r^2), \qquad \dot{\theta} = 1. \tag{5.71}$$

For $\mu < 0$, $r = 0$ when $\dot{r} = 0$; otherwise $\dot{r} < 0$. Thus for all negative μ the phase portraits are attracting spirals. When $\mu = 0$, $\dot{r} = -r^3$ and so the phase portrait is still an attracting spiral. The spiralling trajectories are weakly pitched (the linearized system is a centre when $\mu = 0$). However, for $\mu > 0$, the origin is unstable because $\dot{r} > 0$ when $0 < r < \sqrt{\mu}$. The functions $r(t) \equiv \sqrt{\mu}$, $\theta(t) = t$ are solutions of (5.71) and so the circle $r = \sqrt{\mu}$ forms a closed orbit. For $r > \sqrt{\mu}$, $\dot{r} < 0$ and therefore the closed orbit is a stable limit cycle with trajectories spiralling in from both sides.

The distinct qualitative types of phase portrait are shown in Fig. 5.23 and we conclude that the system (5.70) undergoes a bifurcation at $\mu = 0$. □

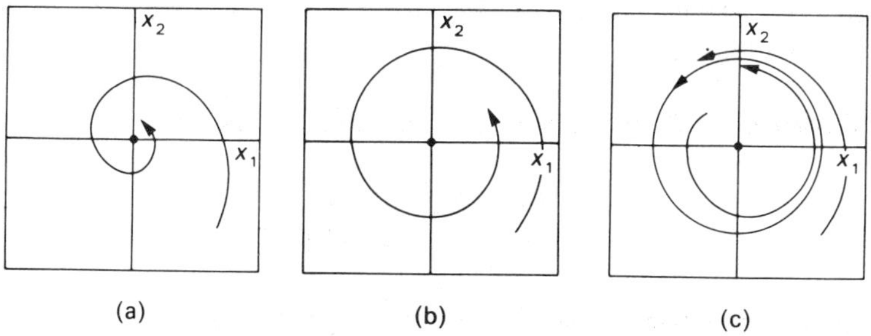

Fig. 5.23. Phase portraits of the parametrized system (5.70): (a) $\mu < 0$; (b) $\mu = 0$; (c) $\mu > 0$.

Observe that the eigenvalues of the linearization of (5.70) at $\mathbf{0}$ are $\mu \pm i$ and become purely imaginary at the bifurcation point $\mu = 0$. For $\mu > 0$ a limit cycle exists which gradually grows in size away from the fixed point as μ increases. This is an example of the *Hopf bifurcation* and we now give sufficient conditions for a limit cycle to appear in this way.

5.5.2 *The Hopf bifurcation*

Theorem 5.5.1. Suppose the parameterized system

$$\dot{x}_1 = X_1(x_1, x_2, \mu), \qquad \dot{x}_2 = X_2(x_1, x_2, \mu) \tag{5.72}$$

has a fixed point at the origin for all values of the real parameter μ. Furthermore suppose the eigenvalues of the linearized system, $\lambda_1(\mu)$ and $\lambda_2(\mu)$, are purely imaginary when $\mu = \mu_0$. If the real part of the eigenvalues $\mathrm{Re}[\lambda_1(\mu)]$ $\{= \mathrm{Re}[\lambda_2(\mu)]\}$, satisfies $(\mathrm{d}/\mathrm{d}\mu)\{\mathrm{Re}[\lambda_1(\mu)]\}|_{\mu=\mu_0} > 0$ and the origin is an asymptotically stable fixed point when $\mu = \mu_0$, then:

(a) $\mu = \mu_0$ is a bifurcation point of the system;
(b) for $\mu \in (\mu_1, \mu_0)$, some $\mu_1 < \mu_0$, the origin is a stable focus;
(c) for $\mu \in (\mu_0, \mu_2)$, some $\mu_2 > \mu_0$, the origin is an unstable focus surrounded by a stable limit cycle, whose size increases with μ.

The Hopf bifurcation is characterized by a stability change of the fixed point accompanied by the creation of a limit cycle. Theorem 5.5.1 gives explicit conditions for such a bifurcation to occur at $\mu = \mu_0$. The mathematics required to prove the theorem is beyond the scope of this book (see Marsden and McCracken, 1976). In contrast, the techniques required to apply the theorem have been discussed already.

Example 5.5.2. Prove that the parameterized system

$$\begin{aligned} \dot{x}_1 &= \mu x_1 - 2x_2 - 2x_1(x_1^2 + x_2^2)^2 \\ \dot{x}_2 &= 2x_1 - \mu x_2 - x_2(x_1^2 + x_2^2)^2, \end{aligned} \tag{5.73}$$

undergoes a Hopf bifurcation at the origin when $\mu = 0$.

Solution. The origin is a fixed point of the system for all μ. The linearized equations are

$$\dot{x}_1 = \mu x_1 - 2x_2, \qquad \dot{x}_2 = 2x_1 + \mu x_2; \tag{5.74}$$

with eigenvalues $\lambda_1(\mu), \lambda_2(\mu) = \mu \pm 2i$ and so when $\mu = 0$, $\lambda_1(0), \lambda_2(0) = \pm 2i$ and $(\mathrm{d}/\mathrm{d}\mu)\{\mathrm{Re}[\lambda_1(\mu)]\}|_{\mu=0} = 1 > 0$ as required.

The system (5.73) with $\mu = 0$ has a strong Liapunov function

$$V(x_1, x_2) = x_1^2 + x_2^2$$

with

$$\dot{V}(x_1, x_2) = -2(2x_1^2 + x_2^2)(x_1^2 + x_2^2)^2 \tag{5.75}$$

so that the origin is asymptotically stable. Therefore by Theorem 5.5.1 the system bifurcates to give a stable limit cycle surrounding the origin for some interval of positive μ. □

When a strong Liapunov function cannot be found, it can be difficult to check that the origin is asymptotically stable, when μ is at the bifurcation point μ_0. The linearization theorem can *never* determine the nature of this non-linear fixed point, because the linearized system is always a *centre*. There is, however, an index which may still determine the stability at $\mu = \mu_0$. This index is calculated by the following procedure:

(a) find the linearization $\dot{\boldsymbol{x}} = \boldsymbol{A}\boldsymbol{x}$ of system (5.72) at the origin of coordinates when $\mu = \mu_0$;

(b) find a non-singular matrix $\boldsymbol{M}$ such that

$$\boldsymbol{M}^{-1}\boldsymbol{A}\boldsymbol{M} = \begin{bmatrix} 0 & |\omega_0| \\ -|\omega_0| & 0 \end{bmatrix}, \tag{5.76}$$

where the eigenvalues of $\boldsymbol{A}$ are $\pm i\omega_0$;

(c) transform the system $\dot{x}_1 = X_1(x_1, x_2, \mu_0)$, $\dot{x}_2 = X_2(x_1, x_2, \mu_0)$ by the change of variables $\boldsymbol{x} = \boldsymbol{M}\boldsymbol{y}$ into $\dot{y}_1 = Y_1(y_1, y_2)$, $\dot{y}_2 = Y_2(y_1, y_2)$;

(d) calculate

$$\begin{aligned} I = |\omega_0|&(Y^1_{111} + Y^1_{122} + Y^2_{112} + Y^2_{222}) \\ &+ (Y^1_{11}Y^2_{11} - Y^1_{11}Y^1_{12} + Y^2_{11}Y^2_{12} \\ &+ Y^2_{22}Y^2_{12} - Y^1_{22}Y^1_{12} - Y^1_{22}Y^2_{22}) \end{aligned} \tag{5.77}$$

where

$$Y^i_{jk} = \frac{\partial^2 Y_i}{\partial y_j \partial y_k}(0, 0) \quad \text{and} \quad Y^i_{jkl} = \frac{\partial^3 Y_i}{\partial y_j \partial y_k \partial y_l}(0, 0).$$

If the index I is *negative*, then the origin is asymptotically stable.

Example 5.5.3. Show that the equation

$$\ddot{x} + (x^2 - \mu)\dot{x} + 2x + x^3 = 0 \tag{5.78}$$

has a bifurcation point at $\mu = 0$ and is oscillatory for some $\mu > 0$.

Solution. The corresponding first-order system is

$$\dot{x}_1 = x_2, \qquad \dot{x}_2 = -(x_1^2 - \mu)x_2 - 2x_1 - x_1^3 \tag{5.79}$$

which has a fixed point at the origin. The eigenvalues of the linearization are given by $\lambda = [\mu \pm \sqrt{(\mu^2 - 8)}]/2$; at $\mu = 0$ they are purely imaginary and $(d/d\mu)[\mathrm{Re}(\lambda)]|_{\mu=0} = \frac{1}{2}$. The linearization of (5.79) is

$$\begin{bmatrix} \dot{x}_1 \\ \dot{x}_2 \end{bmatrix} = \begin{bmatrix} 0 & 1 \\ -2 & \mu \end{bmatrix} \begin{bmatrix} x_1 \\ x_2 \end{bmatrix} \tag{5.80}$$

and so the coefficient matrix is not in the form required to calculate the index I.

The matrix $\boldsymbol{M} = \begin{bmatrix} 1 & 1 \\ 0 & \sqrt{2} \end{bmatrix}$ has the property

$$\boldsymbol{M}^{-1} \begin{bmatrix} 0 & 1 \\ -2 & 0 \end{bmatrix} \boldsymbol{M} = \begin{bmatrix} 0 & \sqrt{2} \\ -\sqrt{2} & 0 \end{bmatrix}$$

as required in (b) above. The change of variable $\boldsymbol{x} = \boldsymbol{M}\boldsymbol{y}$ converts system (5.79) with $\mu = 0$ into

$$\dot{y}_1 = \sqrt{2}y_2, \qquad \dot{y}_2 = -\sqrt{2}y_1 - y_1^2 y_2 - y_1^3/\sqrt{2}, \tag{5.81}$$

and I can now be calculated to be $-2\sqrt{2}$.

Thus as μ increases through 0 system (5.79) bifurcates to a stable limit cycle surrounding an unstable fixed point at the origin. The system (5.79) is the phase plane representation of the equation (5.78) and the existence of a closed orbit implies that $x(t)$ is oscillatory for some $\mu > 0$ □

5.5.3 *An application of the Hopf bifurcation theorem*

Lefever and Nicolis (1971) discuss a simple model of oscillatory phenomena in chemical systems.

They consider the set of chemical reactions

$$\begin{aligned} A &\to X \\ B + X &\to Y + D \\ 2X + Y &\to 3X \\ X &\to E \end{aligned} \tag{5.82}$$

where the inverse reactions are neglected and the initial and final product concentrations A, B, D, E are assumed constant.

The resulting chemical kinetic equations are

$$\dot{X} = a - (b+1)X + X^2Y, \qquad \dot{Y} = bX - X^2Y, \tag{5.83}$$

for some positive parameters a and b. There is a unique fixed point P at $(a, b/a)$. The linearization of (5.83) at P has coefficient matrix

$$\left.\begin{bmatrix} 2XY - b - 1 & X^2 \\ b - 2XY & -X^2 \end{bmatrix}\right|_{(a,\, b/a)} = \begin{bmatrix} b-1 & a^2 \\ -b & -a^2 \end{bmatrix}. \tag{5.84}$$

The determinant of this matrix is a^2 and so the stability of P is determined by the trace. The fixed point is stable for $a^2 + 1 > b$ and unstable for $a^2 + 1 < b$.

To use the Hopf bifurcation theorem we introduce local coordinates $x_1 = X - a$, $x_2 = Y - b/a$ to obtain

$$\begin{aligned} \dot{x}_1 &= (b-1)x_1 + a^2x_2 + 2ax_1x_2 + \frac{b}{a}x_1^2 + x_1^2x_2 \\ \dot{x}_2 &= -bx_1 - a^2x_2 - 2ax_1x_2 - \frac{b}{a}x_1^2 - x_1^2x_2. \end{aligned} \tag{5.85}$$

We now interpret (5.85) as a system parameterized by b with a remaining fixed. The real part of the eigenvalues is $\frac{1}{2}(b - a^2 - 1)$ for $(a-1)^2 < b < (a+1)^2$ and so

$$\frac{\mathrm{d}}{\mathrm{d}b}\left[\tfrac{1}{2}(b - a^2 - 1)\right] = \tfrac{1}{2}$$

at the bifurcation value $b = a^2 + 1$.

All that needs to be checked now is the stability of (5.85) when $b = a^2 + 1$. The matrix

$$\boldsymbol{M} = \begin{bmatrix} a^2 & 0 \\ -a^2 & a \end{bmatrix}$$

satisfies

$$M^{-1}\begin{bmatrix} b-1 & a^2 \\ -b & -a^2 \end{bmatrix} M = \begin{bmatrix} 0 & a \\ -a & 0 \end{bmatrix} \tag{5.86}$$

and the transformation $\boldsymbol{x} = \boldsymbol{M}\boldsymbol{y}$ converts (5.85) into

$$\begin{aligned} \dot{y}_1 &= ay_2 + (1-a^2)ay_1^2 + 2a^2 y_1 y_2 - a^4 y_1^3 + a^3 y_1^2 y_2, \\ \dot{y}_2 &= -ay_1. \end{aligned} \tag{5.87}$$

The stability index (5.77) can now be calculated and, since only Y^1_{111} and $Y^1_{11}Y^1_{12}$ are non-zero, $I = -2a^5 - 4a^3$. It follows that system (5.83) bifurcates to an attracting limit cycle surrounding P as b increases through the critical value $1 + a^2$. An example of a typical phase portrait is given in Fig. 5.24.

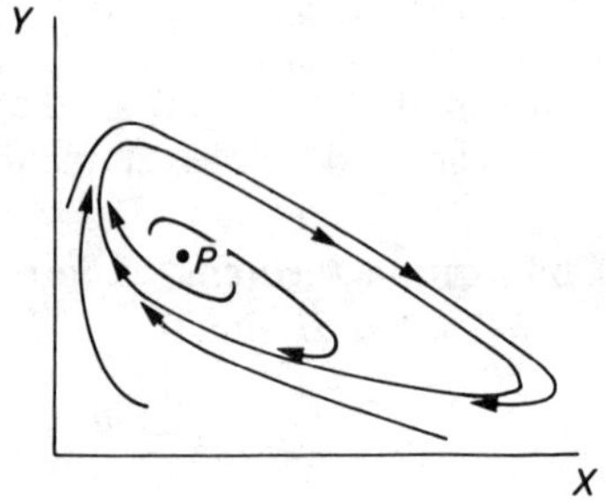

Fig. 5.24. The limit cycle of system (5.83) when $a = 1$, $b = 3$.

5.6 A mathematical model of tumor growth

In recent years the bifurcation of differential equations has attracted much theoretical interest and provided new scope in mathematical models. In this section, we consider a model which illustrates this point particularly well. It concerns the way in which an animal's immune system responds to foreign tissue, in this case a tumor. The model (Rescigno and De Lisi, 1977) is described at length in a survey by Swan (1977); we only sketch the background here.

5.6.1 *Construction of the model*

Tumor cells contain substances (antigens) which cause an immune response in the host animal. This consists of the production of cells (lymphocytes) which attack and destroy the tumor cells.

The following variables are involved in the model (in each case they refer to the size of the cell population described):

(a) L – free lymphocytes on the tumor surface;
(b) C – tumor cells in and on the tumor;
(c) C_S – tumor cells on the surface of the tumor;
(d) $\overline{C}$ – tumor cells on the surface of the tumor not bound by lymphocytes;
(e) C_f – tumor cells in and on the tumor not bound by lymphocytes.

It follows immediately from these definitions that

$$C = C_f - \overline{C} + C_S. \tag{5.88}$$

The tumor is assumed to be spherical at all times, so that

$$C_S = K_1 C^{2/3}, \tag{5.89}$$

where K_1 is constant, and interactions take place only on the surface of the tumor. Not all tumor cells are susceptible to attack and destruction by lymphocytes and only a proportion of free tumor cell-free lymphocyte interactions result in binding. An equilibrium relation

$$C_S - \overline{C} = K_2 \overline{C} L, \tag{5.90}$$

where K_2 is constant, is assumed between the numbers of free and bound lymphocytes, so that (5.88) and (5.89) imply that

$$C_f = C - K_1 K_2 L C^{2/3}/(1 + K_2 L) \tag{5.91}$$

and

$$\overline{C} = K_1 C^{2/3}/(1 + K_2 L). \tag{5.92}$$

This means that L and C can be taken as the basic variables of the model.

The specific growth rate, $\dot{L}/L$, of the free lymphocyte population is assumed to consist of two terms:

(a) a constant death rate λ_1;
(b) a stimulation rate $\alpha_1'\overline{C}(1 - L/L_M)$.

Item (b) shows that, while for small L the stimulation of free lymphocytes increases linearly with $\overline{C}$, there is a maximum population L_M at which the stimulation rate becomes zero. Thus, L satisfies

$$\dot{L} = -\lambda_1 L + \alpha_1'\overline{C}L(1 - L/L_M). \tag{5.93}$$

The growth rate of the tumor cell population C is given by

$$\dot{C} = \lambda_2 C_f - \alpha_2'\overline{C}L. \tag{5.94}$$

The first term in (5.94) describes the growth of tumor cells unaffected by lymphocytes, while the second takes account of free tumor cell-free lymphocyte interactions on the tumor surface.

On substituting for $\overline{C}$ and C_f from (5.91) and (5.92), equations (5.93) and (5.94) can be written as

$$\begin{aligned} \dot{x} &= -\lambda_1 x + \alpha_1 x y^{2/3}\left(1 - \frac{x}{c}\right)/(1 + x) \\ \dot{y} &= \lambda_2 y - \alpha_2 x y^{2/3}/(1 + x), \end{aligned} \tag{5.95}$$

where

$$x = K_2 L, \qquad c = K_2 L_M, \qquad y = K_2 C$$

and $\lambda_1, \lambda_2, \alpha_1, \alpha_2$ are positive parameters. Since x and y are populations they must be non-negative, but x cannot exceed c because L is bounded above by L_M.

We now turn to the qualitative implications of the dynamical equations (5.95) and characterize the various phase portraits that can occur, together with their associated parameter ranges.

5.6.2 *An analysis of the dynamics*

It can be seen at once that the system (5.95) has a saddle point at the origin, for all values of the parameters, by the linearization theorem. The positive x- and y-axes are trajectories of the system and form the separatrices of the saddle.

To investigate the non-trivial fixed points, however, we write (5.95) in the form

$$\dot{x} = xf(x, y), \qquad \dot{y} = y^{2/3}g(x, y), \tag{5.96}$$

where

$$f(x, y) = -\lambda_1 + \alpha_1 y^{2/3}(1 - \frac{x}{c})/(1 + x) \tag{5.97}$$

and

$$g(x, y) = \lambda_2 y^{1/3} - \alpha_2 x/(1 + x). \tag{5.98}$$

The fixed point equations

$$f(x, y) = 0, \qquad g(x, y) = 0$$

give

$$y^{2/3} = \frac{\lambda_1}{\alpha_1}\left(\frac{1+x}{1-(x/c)}\right) = \left(\frac{\alpha_2}{\lambda_2}\frac{x}{1+x}\right)^2,$$

so that the x-coordinates of the non-trivial fixed points satisfy

$$\psi(x) = \frac{x^2(1-(x/c))}{(1+x)^3} = \frac{\lambda_1\lambda_2^2}{\alpha_1\alpha_2^2}. \tag{5.99}$$

The function $\psi(x)$ has a unique global maximum at $x^* = 2c/(c+3)$, with $\psi(x^*) = 4c^2/27(c+1)^2$. Now define

$$\mu_1 = \frac{\lambda_1\lambda_2^2}{\alpha_1\alpha_2^2} - \frac{4c^2}{27(c+1)^2}, \tag{5.100}$$

when it follows that (5.99) has:

(a) *no* real roots for $\mu_1 > 0$;
(b) exactly *one* root at $x = x^*$ for $\mu_1 = 0$; and
(c) exactly *two* roots, x_1^* and x_2^*, where $0 < x_1^* < 2c/(c+3) < x_2^* < c$, when $\mu_1 < 0$.

Geometrically, (a)–(c) correspond to the curves $f(x, y) = 0$ and $g(x, y) = 0$ having the intersections shown in Fig. 5.25.

We now consider the phase portraits for each of these situations.

(*a*) $\mu_1 > 0$

There are no non-trivial fixed points. Recalling that the origin is a saddle point, the signs of $\dot{x}$ and $\dot{y}$ are sufficient to construct a sketch of the phase portrait (see Fig. 5.26).

Observe that $y \to \infty$ as $t \to \infty$, for *all* initial states of the populations. This corresponds to uncontrolled growth of the tumor.

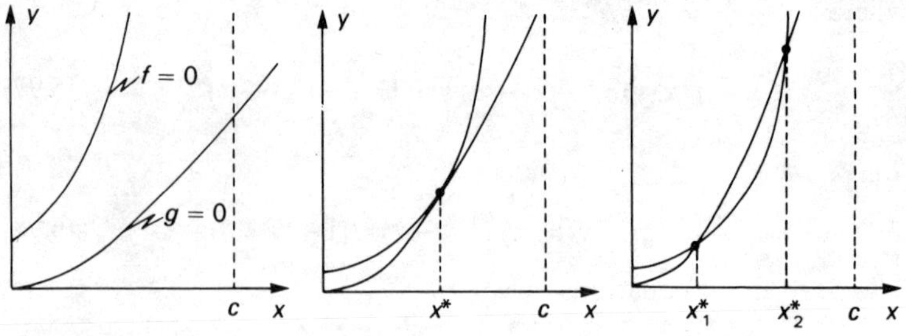

Fig. 5.25. The equation $f(x, y) = 0$ defines a curve which is concave and has asymptote $x = c$, while the curve $g(x, y) = 0$ has asymptote $y = (\alpha_2/\lambda_2)^3$. The possible configurations of these curves are shown: (a) $\mu_1 > 0$; (b) $\mu_1 = 0$; (c) $\mu_1 < 0$.

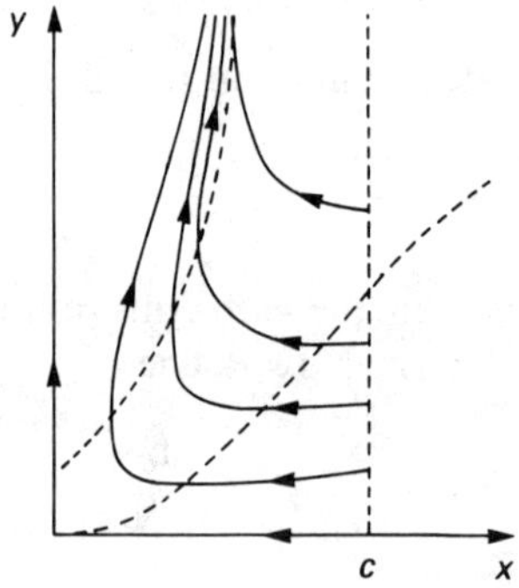

Fig. 5.26. Phase portrait for (5.95) with $\mu_1 > 0$. All trajectories approach $x = c$ asymptotically as $t \to \infty$.

(b) $\mu_1 = 0$

There is a single non-trivial fixed point (x^*, y^*), where $x^* = 2c/(c+3)$. The coefficient matrix of the linearized system is

$$\boldsymbol{W} = \left.\begin{bmatrix} xf_x & xf_y \\ y^{2/3}g_x & y^{2/3}g_y \end{bmatrix}\right|_{(x^*, y^*)}, \tag{5.101}$$

since $f(x^*, y^*) = g(x^*, y^*) = 0$. We have used the notation

$$f_x \equiv \frac{\partial f}{\partial x}, \quad f_y \equiv \frac{\partial f}{\partial y}, \quad g_x \equiv \frac{\partial g}{\partial x} \quad \text{and} \quad g_y \equiv \frac{\partial g}{\partial y}.$$

It follows that

$$\det(\boldsymbol{W}) = x^* y^{*2/3}(f_x g_y - f_y g_x)\big|_{(x^*,\, y^*)}. \tag{5.102}$$

However, as Fig. 5.25(b) shows, when $\mu_1 = 0$ the slopes of the curves $f(x, y) = 0$ and $g(x, y) = 0$ are the same at (x^*, y^*). This implies that

$$\left.\frac{f_x}{f_y}\right|_{(x^*,\, y^*)} = \left.\frac{g_x}{g_y}\right|_{(x^*,\, y^*)} \tag{5.103}$$

so that $\det(\boldsymbol{W}) = 0$. Thus (x^*, y^*) is a non-simple fixed point and linear analysis cannot determine its nature.

We will return to the problem of the precise nature of the fixed point (x^*, y^*) later on. However, the global behaviour of the phase portrait for $\mu_1 = 0$ is clearly such that uncontrolled tumor growth occurs for the majority of initial states (the area of the domain of stability of (x^*, y^*) is finite).

(c) $\mu_1 < 0$

In this case, there are two non-trivial fixed points, $P_1 = (x_1^*, y_1^*)$ and $P_2 = (x_2^*, y_2^*)$ with $0 < x_1^* < x^* < x_2^* < c$ and $x^* = 2c/(c+3)$.

If $\boldsymbol{W}_i\,(i = 1, 2)$ is the coefficient matrix of the linearization at (x_i^*, y_i^*), then (5.97), (5.98) and (5.102) give

$$\det(\boldsymbol{W}_i) = \left.\left\{\alpha_1 \lambda_2 x y^{2/3}\left(\frac{2c}{x} - c - 3\right)\Big/ 3c\,(1 + x)^2\right\}\right|_{(x_i^*,\, y_i^*)} \tag{5.104}$$

Thus, $\det(\boldsymbol{W}_1)$ is positive and $\det(\boldsymbol{W}_2)$ is negative. We can immediately conclude that P_2 is a saddle point, but the stability of P_1 is determined by $\mathrm{tr}(\boldsymbol{W}_1)$. The matrix (5.101), evaluated at (x_1^*, y_1^*), gives

$$\begin{aligned}\mathrm{tr}(\boldsymbol{W}_1) &= (x f_x + y^{2/3} g_y)\big|_{(x_1^*,\, y_1^*)}\\ &= \left.\frac{\lambda_2}{3}\left\{1 - 3\left(\frac{\lambda_1}{\lambda_2}\right)\left(\frac{1+c}{c}\right)\frac{x}{(1+x)(1-x/c)}\right\}\right|_{x_1^*},\\ &= \eta\,(x_1^*)\end{aligned} \tag{5.105}$$

where (5.97) has been used to eliminate y. The function $x\{(1+x)(1-x/c)\}^{-1}$ is strictly increasing on $(0, c)$ and therefore, $\eta(x)$ is a strictly decreasing function of x on $(0, x^*)$ (recall $0 < x^* < c$). There are several possibilities for $\mathrm{tr}(\boldsymbol{W}_1) = \eta(x_1^*)$ as shown in Fig. 5.27.

Observe that $\eta\,(x^*) \geqslant 0$ implies that $\eta\,(x_1^*) > 0$, for any x_1^*, (curves (a) and (b)) whereas if $\eta\,(x^*) < 0$ then $\eta\,(x_1^*)$ may be positive (curve (c)),

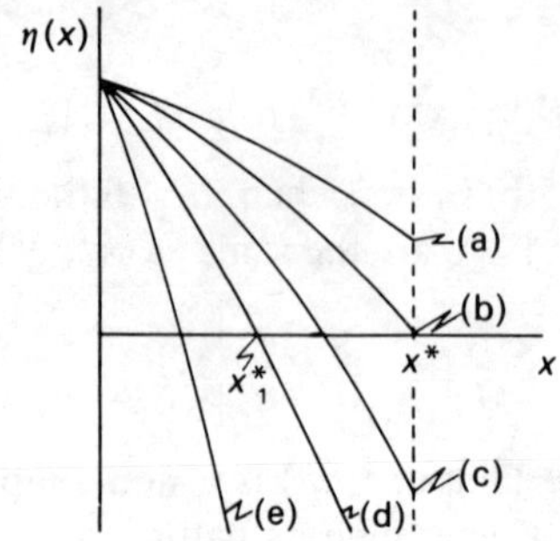

Fig. 5.27. Possible forms for $\eta(x)$ versus x. Observe $\text{tr}(W_1) = \eta(x_1^*)$ is positive for (a)–(c), zero for (d) and negative for (e).

zero (curve (d)) or negative (curve (e)). We can express this result in terms of the parameters of the model in the following way. When $x = x^* = 2c/(c+3)$,

$$\eta(x^*) = \frac{\lambda_2}{3}\left\{1 - \left(\frac{\lambda_1}{\lambda_2}\right)\left(\frac{2(c+3)}{(c+1)}\right)\right\} \tag{5.106}$$

which is zero (i.e. curve (b) in Fig. 5.27) if $\lambda_1/\lambda_2 = (c+1)/2(c+3)$. Now define

$$\mu_2 = \frac{\lambda_1}{\lambda_2} - \frac{(c+1)}{2(c+3)}$$

when it follows that if:

(a) $\mu_2 \leqslant 0$ then $\eta(x^*) \geqslant 0$ and $\text{tr}(\boldsymbol{W}_1) = \eta(x_1^*) > 0$ and hence P_1 is unstable;

(b) $\mu_2 > 0$ then sign of $\text{tr}(\boldsymbol{W}_1)$ is not determined.

When $\mu_2 > 0$, we can say that $\eta(x)$ moves through the sequence of curves (c), (d) and (e) in Fig. 5.27 as μ_2 increases from zero. Correspondingly, the sign of $\text{tr}(\boldsymbol{W}_1) = \eta(x_1^*)$ is positive, zero and then negative. Thus, as μ_2 increases through $(0, \infty)$ (with μ_1 constant) the fixed point P_1 changes from being unstable at $\mu_2 = 0$ to being stable at sufficiently large μ_2.

We summarize the results obtained above in Fig. 5.28. The $\mu_1\mu_2$-plane is shown, divided into three major regions: A, where the only fixed point is at the origin; B, where P_1 is unstable; and C, where

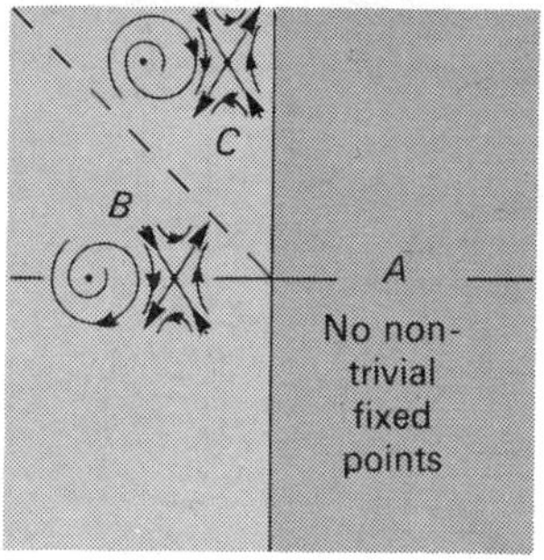

Fig. 5.28. Summary of results of linear stability analysis of (5.95). The nature of the fixed points on $\mu_1 = 0$ and the boundary between B and C are not revealed by this analysis.

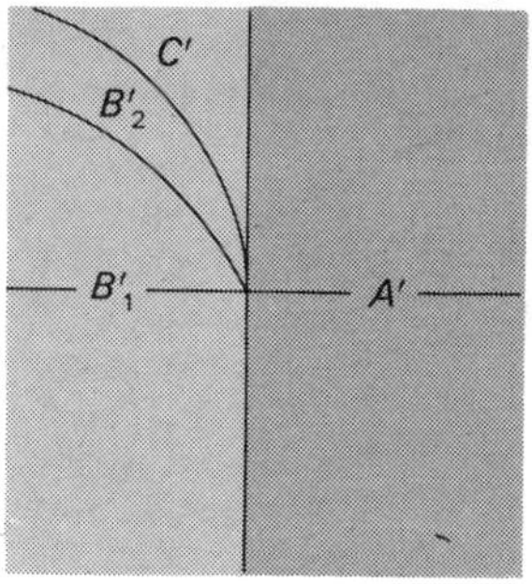

Fig. 5.29. Summary of global analysis of (5.107). The $\bar{\mu}_1\bar{\mu}_2$-plane is divided into four regions A', B'_1, B'_2, C' in which the phase portraits are as shown in Fig. 5.30.

P_1 is stable. We have not determined the shape of the boundary between B and C in the figure and so it is shown schematically as a broken straight line.

The local behaviour shown in Fig. 5.28 is qualitatively equivalent to that of the simpler two parameter system,

$$\dot{x} = -(\bar{\mu}_1 + y^2), \qquad \dot{y} = -(x + \bar{\mu}_2 y + y^2), \tag{5.107}$$

investigated by Takens (1974). The major features of the *global* analysis of (5.107), are summarized in Figs. 5.29 and 5.30. Observe that the local phase portraits at the fixed points are the same in Fig. 5.30(b) and (c). Thus, from the point of view of local behaviour B'_1 and B'_2 are equivalent and we can identify A with A', B with $B' = B'_1 \cup B'_2$ and C with C'.

The relationship between (5.95) and (5.107) extends beyond local behaviour; in fact, every global phase portrait presented by Swan (1977) for (5.95) has a qualitatively equivalent counterpart in Takens' treatment of (5.107). Of course, this does not prove that (5.95) and (5.107) are qualitatively equivalent, since Swan's treatment does not exhaust the global phase portraits of (5.95). However, such an equivalence is certainly a real possibility.

The conjecture that (5.95) and (5.107) are qualitatively equivalent has important repercussions for (5.95) itself. Takens has given a *complete* global analysis of (5.107) and this could be used as a guide in the analysis of (5.95). For example, we would suggest that the

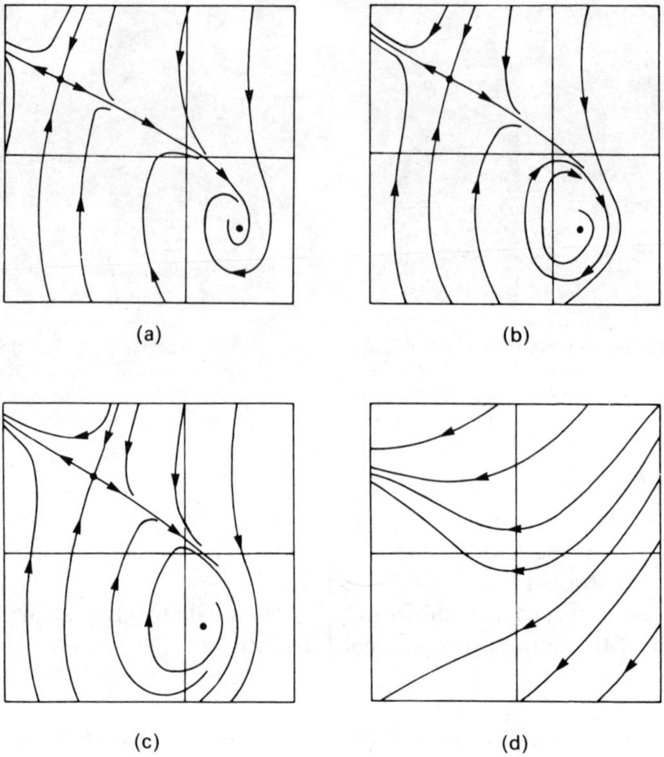

Fig. 5.30. Phase portraits for (5.107) when $(\bar{\mu}_1, \bar{\mu}_2)$ belongs to: (a) C'; (b) B'_2; (c) B'_1; (d) A'.

boundary between C and B in Fig. 5.28 is a Hopf bifurcation, as is the boundary between C' and B' in Fig. 5.29. Remember, the linearization of (5.95) has purely imaginary eigenvalues on the CB-boundary $(\mathrm{tr}(\boldsymbol{W}_1) = 0)$ and there is a sharp change in stability of P_1. These are the 'symptoms' of a Hopf bifurcation. A limit cycle consistent with this surmise appears in Swan's treatment.

There is no mention in Swan's work of the bifurcation corresponding to the $B'_2B'_1$-boundary in Fig. 5.29, but the conjecture of qualitative equivalence would imply its existence. As μ_2 decreases, at fixed $\mu_1 < 0$, we would expect a limit cycle to be created at the BC-boundary by a Hopf bifurcation and then to expand until it reaches the saddle point P_2. A further bifurcation would then occur and the bounded oscillations, arising from initial states within the cycle, would turn into uncontrolled growth of the tumor (see Fig. 5.30(c)).

Finally, qualitative equivalence of (5.95) and (5.107) would allow us to discuss the nature of the fixed points of (5.95) at $\mu_1 = 0$. Takens' classification contains all such 'degenerate' cases for (5.107). For example, when $\bar{\mu}_1 = \bar{\mu}_2 = 0$, the non-simple fixed point of (5.107) has the phase portrait shown in Fig. 5.31. This is qualitatively equivalent to the phase portrait for the non-trivial fixed point (x^*, y^*) of (5.95) given by Swan.

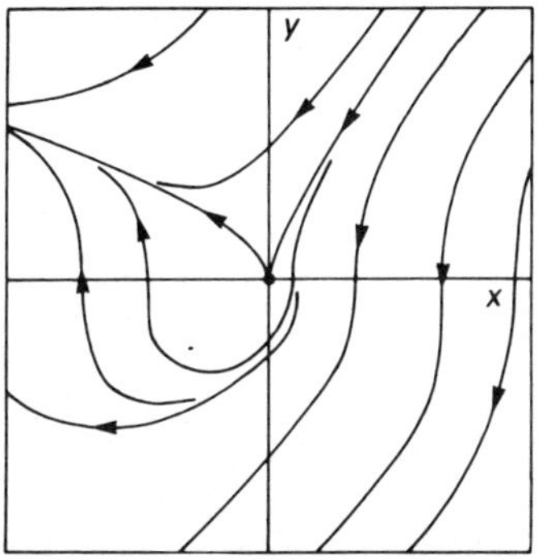

Fig. 5.31. The non-simple fixed point of (5.107) with $\bar{\mu}_1 = \bar{\mu}_2 = 0$.

The four phase portraits shown in Fig. 5.30 are all found for (5.95), at corresponding values of (μ_1, μ_2), and they form the basis of Swan's discussion. As can be seen, uncontrolled growth is the dominant feature. In the parameter space, pairs (μ_1, μ_2) belonging to A and much of B give rise to phase portraits like those in Fig. 5.30(d) and (c) in which almost every trajectory corresponds ultimately to uncontrolled growth. For the remaining set of (μ_1, μ_2), where phase portraits like Fig. 5.30(a) and (b) arise, uncontrolled growth still occurs, unless the initial state of the populations belongs to limited areas of the xy-plane.

A key idea is, then, to find realistic ways in which these domains of stability can be enlarged. For example, Rescigno and De Lisi (1977) consider the existence of a source of lymphocytes entering the system at a constant rate. The dynamical equations (5.95) are replaced by

$$\begin{aligned}\dot{x} &= -\lambda_1(x-x_0)+\alpha_1 x y^{2/3}\left(1-\frac{x}{c}\right)/(1+x)\\ \dot{y} &= \lambda_2 y-\alpha_2 y^{2/3}x/(1+x),\end{aligned} \tag{5.108}$$

where the term $\lambda_1 x_0$, $x_0 > 0$, corresponds to the lymphocyte source. The system (5.108) has a stable fixed point at $(x_0, 0)$, which corresponds to complete remission of the tumor. One of the possible phase portraits is shown in Fig. 5.32. The size of the domain of stability of the fixed point $(x_0, 0)$ increases as the quantity $\lambda_1\lambda_2^2/\alpha_1\alpha_2^2$ decreases. We refer the reader to Swan's (1977) excellent review for further details.

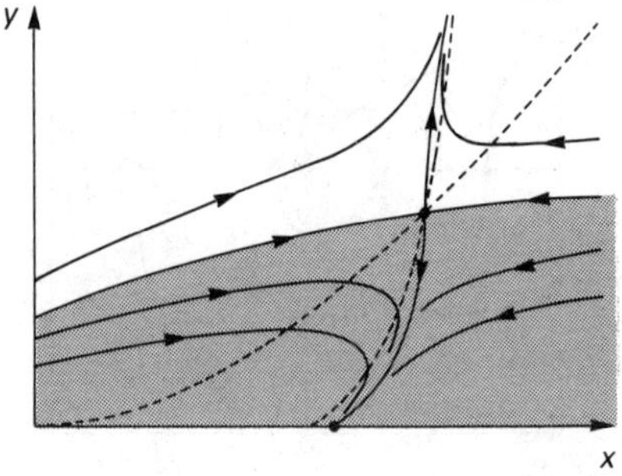

Fig. 5.32. A phase portrait for the lymphocyte source model (5.108). The domain of stability of $(x_0, 0)$ is shaded.

5.7 Exercises

Section 5.4

1. Show that $V(x_1, x_2) = x_1^2 + x_2^2$ is a strong Liapunov function at the origin for each of the following systems:

(a) $\dot{x}_1 = -x_2 - x_1^3, \quad \dot{x}_2 = x_1 - x_2^3;$

(b) $\dot{x}_1 = -x_1^3 + x_2 \sin x_1, \quad \dot{x}_2 = -x_2 - x_1^2 x_2 - x_1 \sin x_1;$

(c) $\dot{x}_1 = -x_1 - 2x_2^2, \quad \dot{x}_2 = 2x_1 x_2 - x_2^3;$

(d) $\dot{x}_1 = -x_1 \sin^2 x_1, \quad \dot{x}_2 = -x_2 - x_2^5;$

(e) $\dot{x}_1 = -(1-x_2)x_1, \quad \dot{x}_2 = -(1-x_1)x_2.$

2. Find domains of stability at the origin for each of the systems given in Exercise 5.1.

3. Show that $V(x_1, x_2) = x_1^2 + x_2^2$ is a weak Liapunov function for the following systems at the origin:
(a) $\dot{x}_1 = x_2, \quad \dot{x}_2 = -x_1 - x_2^3(1 - x_1^2)^2$;
(b) $\dot{x}_1 = -x_1 + x_2^2, \quad \dot{x}_2 = -x_1 x_2 - x_1^2$;
(c) $\dot{x}_1 = -x_1^3, \quad \dot{x}_2 = -x_1^2 x_2$;
(d) $\dot{x}_1 = -x_1 + 2x_1 x_2^2, \quad \dot{x}_2 = -x_1^2 x_2^3$.
Which of these systems are asymptotically stable?

4. Prove that if V is a strong Liapunov function for $\dot{\mathbf{x}} = -\mathbf{X}(\mathbf{x})$, in a neighbourhood of the origin, then $\dot{\mathbf{x}} = \mathbf{X}(\mathbf{x})$ has an unstable fixed point at the origin. Use this result to show that the systems:
(a) $\dot{x}_1 = x_1^3, \quad \dot{x}_2 = x_2^3$;
(b) $\dot{x}_1 = \sin x_1, \quad \dot{x}_2 = \sin x_2$;
(c) $\dot{x}_1 = -x_1^3 + 2x_1^2 \sin x_1, \quad \dot{x}_2 = x_2 \sin^2 x_2$;
are unstable at the origin.

5. Prove that the differential equations
(a) $\ddot{x} + \dot{x} - \dot{x}^3/3 + x = 0$; (b) $\ddot{x} + \dot{x}\sin(\dot{x}^2) + x = 0$;
(c) $\ddot{x} + \dot{x} + x^3 = 0$; (d) $\ddot{x} + \dot{x}^3 + x^3 = 0$,
have asymptotically stable zero solutions $x(t) \equiv 0$.

6. Prove that $V(x_1, x_2) = ax_1^2 + 2bx_1x_2 + cx_2^2$ is positive definite if and only if $a > 0$ and $ac > b^2$. Hence or otherwise prove that

$$\dot{x}_1 = x_2, \qquad \dot{x}_2 = -x_1 - x_2 - (x_1 + 2x_2)(x_2^2 - 1)$$

is asymptotically stable at the origin by considering the region $|x_2| < 1$. Find a domain of stability.

7. Find domains of stability for the following systems by using the appropriate Liapunov function:
(a) $\dot{x}_1 = x_2 - x_1(1 - x_1^2 - x_2^2)(x_1^2 + x_2^2 + 1)$
$\dot{x}_2 = -x_1 - x_2(1 - x_1^2 - x_2^2)(x_1^2 + x_2^2 + 1)$;
(b) $\dot{x}_1 = x_2, \quad \dot{x}_2 = -x_2 + x_2^3 - x_1^5$.

8. Use $V(x_1, x_2) = (x_1/a)^2 + (x_2/b)^2$ to show that the system

$$\dot{x}_1 = x_1(x_1 - a), \qquad \dot{x}_2 = x_2(x_2 - b), \qquad a, b > 0,$$

has an asymptotically stable origin. Show that all trajectories tend to the origin as $t \to \infty$ in the region

$$\frac{x_1^2}{a^2} + \frac{x_2^2}{b^2} < 1.$$

9. Given the system

$$\dot{x}_1 = x_2, \qquad \dot{x}_2 = x_2 - x_1^3$$

show that a positive definite function of the form

$$V(x_1, x_2) = ax_1^4 + bx_1^2 + cx_1x_2 + dx_2^2$$

can be chosen such that $\dot{V}(x_1, x_2)$ is also positive definite. Hence deduce that the origin is unstable.

10. Show that the origin of the system

$$\dot{x}_1 = x_2^2 - x_1^2, \qquad \dot{x}_2 = 2x_1x_2$$

is unstable by using

$$V(x_1, x_2) = 3x_1x_2^2 - x_1^3.$$

11. Show that the fixed point at the origin of the system

$$\dot{x}_1 = x_1^4, \qquad \dot{x}_2 = 2x_1^2x_2^2 - x_2^2$$

is unstable by using the function

$$V(x_1, x_2) = \alpha x_1 + \beta x_2$$

for a suitable choice of the constants α and β.

Verify the instability at the fixed point by examining the behaviour of the separatrices.

Section 5.5

12. State the distinct qualitative types of phase portraits obtained as the parameter μ varies from $-\infty$ to $+\infty$ in the systems:

(a) $\dot{r} = -r^2(r+\mu), \quad \dot{\theta} = 1;$ (b) $\dot{r} = \mu r(r+\mu)^2, \quad \dot{\theta} = 1;$
(c) $\dot{r} = r(\mu - r)(\mu - 2r), \quad \dot{\theta} = 1;$ (d) $\dot{r} = r(\mu - r^2), \quad \dot{\theta} = 1;$
(e) $\dot{r} = r^2\mu, \quad \dot{\theta} = \mu;$ (f) $\dot{r} = r^2, \quad \dot{\theta} = 1 - \mu^2.$

13. Use the stability index to deduce that the following systems are asymptotically stable at the origin:

(a) $\dot{x}_1 = x_2 - x_1^3 + x_1 x_2^2, \quad \dot{x}_2 = -x_1 - x_1 x_2^2;$

(b) $\dot{x}_1 = x_2 - x_1^2 \sin x_1, \quad \dot{x}_2 = -x_1 + x_1 x_2 + 2x_1^2;$

(c) $\dot{x}_1 = x_2 - x_1^2 + 2x_1 x_2 + x_2^2, \quad \dot{x}_2 = -x_1 + x_1 x_2 + x_2^2.$

14. Prove that all the following one-parameter systems undergo Hopf bifurcations at $\mu = 0$ such that a stable limit cycle surrounds the origin for $\mu > 0$:

(a) $\dot{x}_1 = x_2 - x_1^3, \quad \dot{x}_2 = -x_1 + \mu x_2 - x_1^2 x_2;$

(b) $\dot{x}_1 = \mu x_1 + x_2 - x_1^3 \cos x_1, \quad \dot{x}_2 = -x_1 + \mu x_2;$

(c) $\dot{x}_1 = \mu x_1 + x_2 + \mu x_1^2 - x_1^2 - x_1 x_2^2, \quad \dot{x}_2 = -x_1 + x_2^2.$

15. Show that Rayleigh's equation

$$\ddot{x} + \dot{x}^3 - \mu\dot{x} + x = 0$$

undergoes a Hopf bifurcation at $\mu = 0$. Describe the phase portraits near and at $\mu = 0$.

16. Prove that the fixed point at the origin of the system

$$\begin{aligned} \dot{x}_1 &= (\mu - 3)x_1 + (5 + 2\mu)x_2 - 2(x_1 - x_2)^3 \\ \dot{x}_2 &= -2x_1 + (3 + \mu)x_2 - (x_1 - x_2)^3 \end{aligned}$$

has purely imaginary eigenvalues when $\mu = 0$. Find new coordinates y_1, y_2 so that when $\mu = 0$ the linearized part of the system has the correct form for checking stability. Hence or otherwise show that the system undergoes a Hopf bifurcation to stable limit cycles as μ increases through 0.

17. Prove that the one-parameter system

$$\dot{x}_1 = -\mu x_1 - x_2, \qquad \dot{x}_2 = x_1 + x_2^3$$

undergoes a Hopf bifurcation at $\mu = 0$ to unstable limit cycles surrounding a stable focus for $\mu > 0$.

Bibliography

Books

Andronov, A. A. and Chaikin, C. E. (1949), *Theory of Oscillations*, Princeton University Press, Princeton, NJ.

Arnold, V. I. (1973), *Ordinary Differential Equations*, M.I.T. Press, Cambridge, MA.

Arnold, V. I. (1978), *Mathematical Methods of Classical Mechanics*, Graduate Texts in Mathematics, Vol. 60, Springer-Verlag, New York, NY.

Barnett, S. (1975), *Introduction to Mathematical Control Theory*, O.U.P., Oxford.

Braun, M. (1975), *Differential Equations and Their Applications: an Introduction to Applied Mathematics*, Applied Mathematical Sciences, Vol. 15, Springer-Verlag, New York, NY.

Haberman, R. (1977), *Mathematical Models*, Prentice-Hall, Englewood Cliffs, NJ.

Hartman, P. (1964), *Ordinary Differential Equations*, Wiley, New York, NY.

Hartley, B. and Hawkes, T. O. (1970), *Rings, Modules and Linear Algebra*, Chapman and Hall, London.

Hirsch, M. W. and Smale, S. (1974), *Differential Equations, Dynamical Systems and Linear Algebra*, Academic Press, London.

Hayes, P. (1975), *Mathematical Methods in the Social and Managerial Sciences*, Wiley, New York, NY.

Jordan, D. W. and Smith, P. (1977), *Non-linear Ordinary Differential Equations*, Clarendon Press, Oxford.

Marsden, J. E. and McCracken, M. (1976), *The Hopf Bifurcation and its Applications*, Applied Mathematical Sciences, Vol. 19, Springer-Verlag, New York, NY.

May, R. M. (1974), *Stability and Complexity in Model Ecosystems*, Monographs in Population Biology, Princeton University Press, Princeton, NJ.

Maynard-Smith, J. (1974), *Models in Ecology*, C.U.P., London.

Petrovski, I. G. (1966), *Ordinary Differential Equations*, Prentice-Hall, Englewood Cliffs, NJ.

Pieliou, E. C. (1977), *Mathematical Ecology*, Wiley, New York, NY.

Swan, G. W. (1977), *Some Current Mathematical Topics in Cancer Research*, Xerox University Microfilms, Ann Arbor, Michigan, MI.

Takens, F. (1974), *Applications of Global Analysis I*, Comm. Math. Inst. Rijksuniversitiet Utrecht, 3.

Papers

Goodwin, R. M. (1951), The non-linear accelerator and the persistence of business cycles, *Econ.*, **19**, 1–17.

Lefever, R. and Nicolis, G. (1971), Chemical instabilities and sustained oscillations, *J. Theor. Biol.*, **30**, 267–284.

Rescigno, A. and De Lisi, C. (1977), Immune surveillance and neoplasia II, *Bull. Math. Biol.*, **39**, 487–497.

Swan, G. W. (1979), Immunological surveillance and neoplastic development, *Rocky Mountain J. Math.*, **9**, 143–148.

Tanner, J. T. (1968), The stability and intrinsic growth rate of prey and predator populations, *Ecol.*, **56**, 855–867.

Zeeman, E. C. (1973), *Differential Equations for the Heartbeat and Nerve Impulse*, Salvador Symposium on Dynamical Systems, Academic Press, 683–741.

Zeeman, E. C. (1979), *Population Dynamics from Game Theory*, Int. Conf. Global Theory of Dynamical Systems, Northwestern University, Evanston, IL.

Hints to exercises

Chapter 1

1. (a) $x = 1 + t + Ce^t$.
   (b) $x = te^t + Ce^t$.
   (c) $x = -5e^{\cos t}\operatorname{cosec} t + C \operatorname{cosec} t, \quad (n-1)\pi < t < n\pi,$
   n integer.
2. (a) $x = Ce^{t^2/2}, \quad x > 0.$
   $x = C'e^{t^2/2}, \quad x < 0.$
   $x \equiv 0.$
   (b) $x = C/t, \quad t < 0.$
   $x = C'/t, \quad t > 0.$
   (c) $x = \sqrt{(C - t^2)}, \quad -\sqrt{C} < t < \sqrt{C}.$
   $x = -\sqrt{(C' - t^2)}, \quad -\sqrt{C'} < t < \sqrt{C'}.$
   (d) $x = \dfrac{C}{\sinh t}, \quad t < 0.$
   $x = \dfrac{C'}{\sinh t}, \quad t > 0.$
3. $F(t, x) = x \ln(tx).$
4. (a) $x = \dfrac{1}{C - t}, \quad t < C.$
   $x = \dfrac{1}{C' - t}, \quad t > C'.$
   $x \equiv 0.$

(b) $x = (C-t)^{-1/2}, t < C$. $x = -(C'-t)^{-1/2}, t < C'$.
$x \equiv 0$.

(c) $x = (1+Ce^t)/(1-Ce^t)$, $x < -1, t > -\ln C$, $C \in \mathsf{R}^+$.
$x = (1-C'e^t)/(1+C'e^t)$, $-1 < x < 1$, $-\infty < t < \infty$, $C' \in \mathsf{R}^+$.
$x = (1+\overline{C}e^t)/(1-\overline{C}e^t)$, $x > 1$, $t < -\ln \overline{C}$, $\overline{C} \in \mathsf{R}^+$.
$x \equiv -1$. $x \equiv 1$.

(d) $x = (t-C)^3$, $t < C$; $x \equiv 0$, $C \leqslant t \leqslant C'$;
$x = (t-C')^3$, $C' < t$.
$x = (t-C)^3$, $t < C$; $x \equiv 0$, $C \leqslant t$.
$x \equiv 0$, $t < C$; $x = (t-C)^3$, $C \leqslant t$.
$x \equiv 0$.

(e) $x \equiv -1$, $t < C - \pi/2$;
$x = \sin(t-C)$, $C - \pi/2 \leqslant t \leqslant C + \pi/2$;
$x \equiv 1$, $t > C + \pi/2$.
$x \equiv 1$. $x \equiv -1$.

(f) $x \equiv 0$, $t \leqslant C$; $x = (t-C)^2$, $t > C$.
$x \equiv 0$.
Yes for both (e) and (f)

5. 4(d) $x = (t-C)^3$, $t \leqslant C$;
$x \equiv 0$, $t > C$,
satisfies $x(0) = 0$ for all negative C.

4(f) $x \equiv 0$, $t < C$;
$x = (t-C)^2$, $t > C$,
satisfies $x(0) = 0$ for all positive C.

4(e) $x = \begin{cases} -1, & t < C - \pi/2 \\ \sin(t-C), & C - \pi/2 \leqslant t \leqslant C + \pi/2 \\ 1, & t > C + \pi/2 \end{cases}$ satisfies $x(0) = +1$
for $C < -\pi/2$ and $x(0) = -1$ for $C > \pi/2$.

6. (a) $\xi(-t)$ is a solution.
(b) $-\xi(t)$ is a solution.

7. If $C = h(t_0, x_0)$ then $C = h(\alpha t_0, \alpha x_0)$, $\forall \alpha \neq 0$.

Substituting $u = x/t$ in $\dot{x} = e^{x/t}$ results in $\displaystyle\int^x \frac{du}{e^u - u} = \log t + C$;

the integral *cannot* be evaluated using elementary calculus.

8. $\ddot{x} = \dot{x}\dfrac{d}{dx}(\dot{x}) = (a-2bx)x(a-bx)$ gives the regions of concavity and convexity in the tx-plane.

9. $\dot{y} = -ay + b$; $y = \dfrac{b}{a} + Ce^{-at}$.

11. (b), (d) and (e) are autonomous equations; the isoclines are x = constant lines.

12. (a) $1(R)$; (b) $-1(A)$, $0(R)$, $1(A)$; (c) $0(S)$; (d) $-1(A)$, $0(S)$, $2(R)$; (e) none
$[(A) \equiv$ attractor, $(S) \equiv$ shunt, $(R) \equiv$ repellor.$]$

13. $\{$(a), (b), (f)$\}$, $\{$(c), (e)$\}$, $\{$(d)$\}$.

14. $\lambda \leqslant 0$  ; $\lambda > 0$ $(\lambda \neq 1)$ 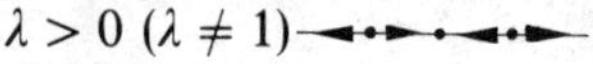;
$\lambda = 1$ .

15. 16; 2^{n+1}.

16. $\dot{y} = y(y-c)$ transforms to $k\dot{x} = (kx+l)(kx+(l-c))$; $k = 1$ and $l - c = -a$, $l = -b$ or $l - c = -b$, $l = -a$.

17. Let $y = x^3 + ax - b$. The numbers of real roots of $y = 0$ are determined by:

$$a > 0 \Rightarrow \frac{dy}{dx} > 0 \Rightarrow \text{one root};$$

$$a < 0 \Rightarrow y_{\max} = -\frac{2a}{3}\sqrt{\left(-\frac{a}{3}\right)} - b,\ y_{\min} = +\frac{2a}{3}\sqrt{\left(-\frac{a}{3}\right)} - b;$$

(i) $y_{\max} < 0 \Rightarrow 1$ root;
(ii) $y_{\min} > 0 \Rightarrow 1$ root;
(iii) $y_{\max} > 0 > y_{\min} \Rightarrow 3$ roots.

18. b/q.

19. (a) $(0, 0)$, $(c/d, a/b)$; (b) $(n\pi, 0)$, n integer; (c) $(0, 0)$; (d) $(0, 0)$, $(2, 0)$, $(0, 2)$, $(2/3, 2/3)$; (e) $(n\pi, (2m+1)\pi/2)$, n, m integers.

20. (a) $\dot{x}_1 = -x_2$, $\dot{x}_2 = x_1$; (b) $\dot{x}_1 = -\frac{1}{2}x_2$, $\dot{x}_2 = 2x_1$;
(c) $\dot{x}_1 = x_1$, $\dot{x}_2 = -2x_2$; (d) $\dot{x}_1 = x_2$, $\dot{x}_2 = x_1$;
(e) $\dot{x}_1 = x_1 + x_2$, $\dot{x}_2 = 2x_2$.

21. $\{$(a), (b)$\}$, $\{$(c), (d)$\}$, $\{$(e)$\}$

22. Both systems satisfy the same differential equation
$$\frac{dx_2}{dx_1} = \frac{X_2(x_1, x_2)}{X_1(x_1, x_2)}.$$

23. $\dot{y} = y_1$, $\dot{y}_2 = -y_2$.

24. Assume another solution $(x_1'(t), x_2'(t))$ and consider the differences $x_1 - x_1'$, $x_2 - x_2'$.

25. See Exercise 1.9.

26. Invariance of the system equations under $\mathbf{x} \to -\mathbf{x}$ means that $-\dot{\mathbf{x}} = \mathbf{X}(-\mathbf{x})$. Thus $\mathbf{X}(\mathbf{x}) = -\mathbf{X}(-\mathbf{x})$ (see Exercise 1.6).

30. (a) The isoclines are all straight lines passing through the point (2, 1). Moreover the slope is always perpendicular to the isocline (closed orbits exist).

(b) Locate the regions of constant $\dot{x}_1$, $\dot{x}_2$ sign and consider trajectory directions (*no* closed orbits).

(c) The isoclines are radial and the tangents to the trajectories make acute angles with outward direction (*no* closed orbits).

31. Slope k-isocline satisfies $kx_2 + x_2^2 = -x_1$. Solutions:
$e^{2x_1}(x_2^2 + x_1 - \frac{1}{2}) = C$.

32. $\dfrac{x^2}{1-x^2} = Ce^{2t}$

33. (a) $\phi_t(x) = \sinh^{-1}(e^t \sinh x)$; (b) $\phi_t(x) = x^{(e^t)}$.

34. (a) $x_1(t) = e^t x_1(0)$, $x_2(t) = \dfrac{e^t x_1(0)}{2} + e^{-t}\left[x_2(0) - \dfrac{x_1(0)}{2}\right]$, all t.

(b) $x_2(t) = \dfrac{1}{[1/x_2(0)] - t}$; substitute in $\dot{x}_1 = x_1 x_2$,

$t < \dfrac{1}{x_2(0)}$, $t > \dfrac{1}{x_2(0)}$.

36. For $x_0 > 0$, $\ln\left(\dfrac{x}{x_0}\right) = \dfrac{t^2}{2} - \dfrac{t_0^2}{2}$ using initial conditions. For $x_0 = 0$, $x \equiv 0$.

Chapter 2

2. $\{(a)\}$, $\{(b), (f)\}$, $\{(c), (d)\}$, $\{e\}$.

3. $J, M = \begin{bmatrix} 3 & 1 \\ 0 & 3 \end{bmatrix}, \begin{bmatrix} 1 & -1 \\ -1 & 2 \end{bmatrix}; \begin{bmatrix} 2 & -2 \\ 2 & 2 \end{bmatrix}, \begin{bmatrix} -1 & 1 \\ 3 & -2 \end{bmatrix};$
$\begin{bmatrix} 3 & 0 \\ 0 & 2 \end{bmatrix}, \begin{bmatrix} -1 & 2 \\ 4 & -7 \end{bmatrix}.$

4. $J, M = \begin{bmatrix} 2 & 0 \\ 0 & 1 \end{bmatrix}, \begin{bmatrix} 1 & 1 \\ 2 & 1 \end{bmatrix}; \begin{bmatrix} 0 & -1 \\ 1 & 0 \end{bmatrix}, \begin{bmatrix} 41 & -1 \\ 58 & 0 \end{bmatrix};$
$\begin{bmatrix} 3 & 1 \\ 0 & 3 \end{bmatrix}, \begin{bmatrix} -2 & -1/3 \\ 3 & 0 \end{bmatrix}.$

5. $\dot{y}_1 = -y_1, \dot{y}_2 = -2y_2 - 3y_3, \dot{y}_3 = -2y_2 - 4y_3.$

6. Similar matrices have the same eigenvalues. Check that the Jordan matrices in Proposition 2.2.1 are determined uniquely by their eigenvalues for types (a) and (b). Are $\begin{bmatrix} \lambda & 1 \\ 0 & \lambda \end{bmatrix}$ and $\begin{bmatrix} \lambda & 0 \\ 0 & \lambda \end{bmatrix}$ similar?

8. Use Exercise 2.4 to obtain canonical systems, solve these and use $\boldsymbol{x} = \boldsymbol{My}$.

11. $\dot{x}_1 = -\frac{3}{2}x_1 + \frac{5}{2}x_2, \dot{x}_2 = -\frac{1}{2}x_1 - \frac{1}{2}x_2.$

12. $x_1(t) = (2e^{-2t} - e^{-3t})x_1(0) + (e^{-3t} - e^{-2t})x_2(0);$
$x_2(t) = (2e^{-2t} - 2e^{-3t})x_1(0) + (-e^{-2t} + 2e^{-3t})x_2(0).$
Put $y_1 = 4e^{-2t} = x_1 + 3x_2$. Compare coefficients of e^{-2t} and e^{-3t} and obtain simultaneous equations for $x_1(0)$ and $x_2(0)\,[x_1(0) = x_2(0) = 1]$.

13. (a) unstable node; (b) saddle; (c) centre; (d) stable focus; (e) unstable improper node.

14. $x_1 = y_1 + 3y_2, x_2 = 2y_1 + 4y_2; \dot{y}_1 = -y_1, \dot{y}_2 = -3y_2.$

15. Linear transformations are continuous and so

$$\lim_{t \to \infty} (\boldsymbol{Nx}(t)) = \boldsymbol{N}\left(\lim_{t \to \infty} \boldsymbol{x}(t)\right).$$

$$\lim_{t \to \infty} \boldsymbol{y}(t) = \lim_{t \to \infty} N\boldsymbol{x}(t) = \boldsymbol{N}\left(\lim_{t \to \infty} \boldsymbol{x}(t)\right) = \boldsymbol{NO} = \boldsymbol{0}.$$

If $\dot{\boldsymbol{x}} = \boldsymbol{Ax}$ has a stable fixed point at $\boldsymbol{O}$, then $\lim_{t \to \infty} \mathbf{x}(\mathbf{t}) = \mathbf{0}$ for *all* trajectories $\mathbf{x}(t)$ and so $\lim_{t \to \infty} \mathbf{y}(t) = \mathbf{0}$ for *all* trajectories of $\dot{\boldsymbol{y}} = \boldsymbol{NAN}^{-1}\boldsymbol{y}$.

16. If $\boldsymbol{y}(t) = \boldsymbol{Nx}(t)$, then $\lim_{t \to -\infty} \boldsymbol{y}(t) = \lim_{t \to -\infty} \boldsymbol{Nx}(t) = \boldsymbol{O}$. If $\mathbf{0}$ is a saddle point then there exist two trajectories $\mathbf{x}(t)$ and $\mathbf{x}'(t)$ such that $\lim_{t \to \infty} \mathbf{x}(t) = \lim_{t \to -\infty} \mathbf{x}'(t) = \mathbf{0}$. The corresponding trajectories $\boldsymbol{y}(t) = N\boldsymbol{x}(t)$, $\boldsymbol{y}'(t) = N\boldsymbol{x}'(t)$ satisfy $\lim_{t \to \infty} \mathbf{y}(t) = \lim_{t \to \infty} \mathbf{y}'(t) = \mathbf{0}$. The saddle is the only fixed point of linear systems with this property.

17. The linear transformation must preserve straight lines and so either $y_1 = ax_1$, $y_2 = dx_2$ or $y_1 = bx_2$, $y_2 = cx_1$. Then consider

$$y_2 = C'y_1^{\mu'} \Rightarrow x_2 = \frac{C'a^{\mu'}}{d} x_1^{\mu'} \Rightarrow \mu = \mu';$$

$$y_2 = C'y_1^{\mu'} \Rightarrow x_2 = \frac{C'b^{\mu'}}{c} x_2^{\mu'} \Rightarrow \mu = \frac{1}{\mu'}.$$

Consider $\mu = \lambda_2/\lambda_1$, $\mu' = \nu_2/\nu_1$.

18. If $\lambda_2 k = \lambda_1$, then $\dot{y}_1 = \lambda_1 y_1$, $\dot{y}_2 = \lambda_1 y_2$ which is qualitatively equivalent to $\dot{y}_1 = \varepsilon y_1$, $\dot{y}_2 = \varepsilon y_2$ with $\varepsilon = \operatorname{sgn}(\lambda_1)$.

20. (a) $\begin{bmatrix} 1 & 0 \\ 0 & -1 \end{bmatrix}$; (b) $\begin{bmatrix} -1 & 0 \\ 0 & -1 \end{bmatrix}$; (c) $\begin{bmatrix} 0 & -1 \\ 1 & 0 \end{bmatrix}$;

(d) $\begin{bmatrix} 0 & 1 \\ 1 & 0 \end{bmatrix}$.

21. (a) $\begin{bmatrix} e^{2t} & 0 \\ 0 & e^{3t} \end{bmatrix}$; (b) $\begin{bmatrix} e^t & 2e^t(e^t - 1) \\ 0 & e^{2t} \end{bmatrix}$;

(c) $\dfrac{e^{6t}}{7}\begin{bmatrix} 3 & 4 \\ 3 & 4 \end{bmatrix} + \dfrac{e^{-t}}{7}\begin{bmatrix} 4 & -4 \\ -3 & 3 \end{bmatrix}$;

(d) $\dfrac{e^{-2t}}{\sqrt{2}}\begin{bmatrix} \sqrt{2}\cos\beta t & \sin\beta t \\ -2\sin\beta t & \sqrt{2}\cos\beta t \end{bmatrix}$, $\beta = 2\sqrt{2}$;

(e) $e^{-3t}\begin{bmatrix} 1-t & t \\ -t & 1+t \end{bmatrix}$.

22. (a) $\dfrac{e^{-t}}{4}\begin{bmatrix} 7 & -7 \\ 3 & -3 \end{bmatrix} + \dfrac{e^{-5t}}{4}\begin{bmatrix} -3 & 7 \\ -3 & 7 \end{bmatrix}$;

(b) $e^{4t}\begin{bmatrix} \cos 2t - \sin 2t & 2\sin 2t \\ -\sin 2t & \cos 2t + \sin 2t \end{bmatrix}$.

23. Let $y_1 = x_1$, $y_2 = x_3$, $y_3 = x_2$, $y_4 = x_4$.

24. $\dfrac{(\boldsymbol{P}+\boldsymbol{Q})^k}{k!} = \sum_{l=0}^{k} \dfrac{1}{l!(k-l)!} P^l Q^{k-1}$.

25. (a) $\boldsymbol{C}^2 = \boldsymbol{0}$; $e^{At} = e^{\lambda_0 t}(\boldsymbol{I} + t\boldsymbol{C})$.

(b) $e^{\alpha t\boldsymbol{I}} = e^{\alpha t}\boldsymbol{I}$;

$$e^{\beta t\boldsymbol{D}} = \sum_{n=0}^{\infty} \frac{(-1)^n}{(2n)!}(\beta t)^{2n}\boldsymbol{I} + \sum_{n=0}^{\infty} \frac{(-1)^n}{(2n+1)!}(\beta t)^{2n+2}\boldsymbol{D},$$

$$\boldsymbol{D} = \begin{bmatrix} 0 & -1 \\ 1 & 0 \end{bmatrix}.$$

26. The characteristic equation of $\boldsymbol{A}$ is

$$\lambda^2 - (\lambda_1 + \lambda_2)\lambda + \lambda_1\lambda_2 = 0$$

where λ_1, λ_2 are the eigenvalues. The Cayley–Hamilton Theorem states

$$\boldsymbol{A}^2 - (\lambda_1 + \lambda_2)\boldsymbol{A} + \lambda_1\lambda_2\boldsymbol{I} = \boldsymbol{0} \tag{1}$$

and so

$$(\boldsymbol{A} - \lambda_1\boldsymbol{I})(\boldsymbol{A} - \lambda_2\boldsymbol{I}) = \boldsymbol{0}.$$

Calculate $(\boldsymbol{A} - \lambda_1\boldsymbol{I})^2$ by expanding and then substituting for $\boldsymbol{A}^2$ from (1).

27. Show $e^{\boldsymbol{A}t} = e^{\lambda_0 t}.e^{\boldsymbol{Q}t}$ and $\boldsymbol{Q}^3 = \boldsymbol{0}$. Use (2.63).

$$e^{\boldsymbol{A}t} = e^{\lambda_0 t}\left[I + t\boldsymbol{Q} + \ldots + \frac{t^{n-1}\boldsymbol{Q}^{n-1}}{(n-1)!}\right].$$

28. (a) $y_1 = x_1 + 1$, $y_2 = x_2 + 1$; (d) $y_1 = x_1 + \frac{1}{4}$, $y_2 = x_2 - \frac{1}{2}$;
(e) $y_1 = x_1 - 1/3$, $y_2 = x_2 + 5/3$, $y_3 = x_3 + 1/3$.

29. $\boldsymbol{x}(t) = \{\frac{1}{2}(a-b) + \frac{1}{2}t\}\begin{bmatrix} 1 \\ -1 \end{bmatrix} + \{\frac{1}{2}e^{2t}(a+b) - \frac{1}{4}(1 - e^{2t})\}\begin{bmatrix} 1 \\ 1 \end{bmatrix}.$

30. If $\boldsymbol{x} = \boldsymbol{M}\boldsymbol{y}$, then $\dot{\boldsymbol{x}} = \boldsymbol{A}\boldsymbol{x} + \boldsymbol{h}$ transforms to
$\dot{\boldsymbol{y}}(t) = \boldsymbol{M}^{-1}\boldsymbol{A}\boldsymbol{M}\boldsymbol{y}(t) + \boldsymbol{M}^{-1}\boldsymbol{h}(t)$.
If $\boldsymbol{A}$ has real distinct eigenvalues λ_1, λ_2 choose matrix $\boldsymbol{M}$ such that

$$\boldsymbol{M}^{-1}\boldsymbol{A}\boldsymbol{M} = \begin{bmatrix} \lambda_1 & 0 \\ 0 & \lambda_2 \end{bmatrix}.$$

No; if $\boldsymbol{h}(t) \not\equiv \boldsymbol{0}$ then $\boldsymbol{M}^{-1}\boldsymbol{h}(t) \not\equiv \boldsymbol{0}$.

31. (a) Convert to linear system; (b) use isoclines.

32. For example,

$$\boldsymbol{M} = \begin{bmatrix} 1 & -2 & 0 \\ 0 & -1 & 0 \\ 0 & -1 & 1 \end{bmatrix} \quad \text{and } a = 1, b = 2, c = 3.$$

Unstable focus in the y_1y_2-plane. Repellor on the y_3-axis.

33. $x_1 = 5e^{2t} - (4t+5)e^t$, $x_2 = 2e^{2t} - 2e^t$, $x_3 = e^{2t}$.

34. $e^{At} =$:

(a) $$e^{\lambda t}\begin{bmatrix} 1 & t & \frac{t^2}{2} & \frac{t^3}{6} \\ 0 & 1 & t & \frac{t^2}{2} \\ 0 & 0 & 1 & t \\ 0 & 0 & 0 & 1 \end{bmatrix}$$

(b) $$\left[\begin{array}{c|c} e^{\alpha t}\begin{bmatrix} \cos\beta t & -\sin\beta t \\ \sin\beta t & \cos\beta t \end{bmatrix} & \boldsymbol{0} \\ \hline \boldsymbol{0} & e^{\lambda t}\begin{bmatrix} 1 & t \\ 0 & 1 \end{bmatrix} \end{array}\right].$$

(c) $$\left[\begin{array}{c|c} e^{\alpha t}\begin{bmatrix} \cos\beta t & -\sin\beta t \\ \sin\beta t & \cos\beta t \end{bmatrix} & \boldsymbol{0} \\ \hline \boldsymbol{0} & e^{\gamma t}\begin{bmatrix} \cos\delta t & -\sin\delta t \\ \sin\delta t & \cos\delta t \end{bmatrix} \end{array}\right]$$

35. Subsystems are formed from the following subsets of coordinates: $\{x_1\}$, $\{x_2, x_3\}$, $\{x_4\}$, $\{x_5, x_6\}$ and are respectively a repellor, centre, repellor and unstable improper node.

Chapter 3

2. Let $y_1 = f(r)\cos\theta$, $y_2 = f(r)\sin\theta$. Show the circle $x_1^2 + x_2^2 = r$ maps onto the circle $y_1^2 + y_2^2 = f(r)$. The result illustrates the qualitative equivalence of the global phase portrait and the local phase portrait at the origin.

4. (i) $\dot{y}_1 = y_2$, $\dot{y}_2 = 2y_1 - 3y_2$;
(ii) $\dot{y}_1 = 2y_1 + y_2$, $\dot{y}_2 = y_1$;
(iii) $\dot{y}_1 = 2e^{-\frac{1}{2}}y_1 - e^{-1}y_2$, $\dot{y}_2 = -y_2$.

5. (a) $(1, -1)$, saddle; $(1, 1)$, stable node; $(2, -2)$, unstable focus; $(2, 2)$, saddle.
(b) $(\pm 1, 0)$, saddle; $(0, 0)$, centre;
(c) $(m\pi, 0)$, unstable improper node (m even), saddle (m odd);
(d) $(0, -1)$, stable focus;

(e) (0, 0), non-simple; (2, −2), saddle;
(f) (0, 0), stable focus;
(g) (−1, −1), stable focus; (4, 4), unstable focus.

6. Solve $\dfrac{dx_2}{dx_1} = \dfrac{x_1 - x_1^5}{-x_2}$. The function $f(x_1, x_2) = 3(x_1^2 + x_2^2) - x_1^6$ has a minimum at the origin and so locally the level curves of f are closed. The Linearization Theorem is not applicable because the linearization is a centre.

7. (a) (1, 0); (b) (1, 2), (1, 0); (c) $(\sqrt{2}, 1)$, $(-\sqrt{2}, 1)$.

8. $x_1 \dfrac{du}{dx_1} = u - x_1^3$; $u = Cx_1 - \dfrac{x_1^3}{2}$.

9. Yes; the identity map suffices for qualitative equivalence. Both systems satisfy $\dfrac{dx_2}{dx_1} = \dfrac{x_2}{x_1}$ and both have a unique fixed point at the origin.

10. If a trajectory lies on the line $x_2 = kx_1$, then $\dfrac{dx_2}{dx_1} = k$.

11. The line of fixed points is $x_1^2 = x_2^3$. The linearization at the fixed point (k^3, k^2), k real, is

$$\begin{bmatrix} 2k^3 & -3k^4 \\ 2k^9 & -3k^{10} \end{bmatrix}.$$

Yes, assume the fixed point is simple and use linearization theorem to obtain a contradiction; Yes.

12. Let N be the neighbourhood of a fixed point with boundary given by a closed trajectory. Then $\mathbf{x}_0 \in N$ satisfies $\phi_t(\mathbf{x}_0) \in N$ for all $t \geqslant 0$ and hence the fixed point is stable. Given $\mathbf{x}_0 \in N \backslash \{\mathbf{0}\}$, $\mathbf{x}_0 = \phi_T(\mathbf{x}_0)$ for some positive T, since every trajectory in N is periodic, and so $\phi_t(\mathbf{x}_0) \not\to \mathbf{0}$ as $t \to \infty$.

13. $\dot{y}_1 = \dot{x}_1 + 3x_2^2\dot{x}_2 = 0$, $\dot{y}_2 = \dot{x}_2 + 2x_2\dot{x}_2 = 1$. y_1, y_2 are differentiable functions of x_1 and x_2; inverse transformations

$$x_1 = y_1 - x_2^3 \text{ and } x_2 = \frac{-1 + \sqrt{(1 + 4y_2)}}{2}$$

are differentiable functions of y_1 and y_2 providing $y_2 > -\frac{1}{4}$ (this condition is necessarily satisfied if $y_2 = x_2 + x_2^2$).

16. (−1, 0) saddle; (0, 0) unstable node; (1, 0) saddle. Observe that the lines $x_1 = 0, \pm 1$ and $x_2 = 0$ are all unions of trajectories.

17. (a) $\dfrac{x_1^3}{3}+x_1-\dfrac{x_2^2}{2}$, $D=\mathsf{R}^2$;

(b) $x_1+\ln|x_1|+x_2+\ln|x_2|$, $D=\{(x_1,x_2)|x_1x_2\neq 0\}$;

(c) $\dfrac{1}{x_2}-\sin x_1$, $D=\{(x_1,x_2)||x_1|<\pi/2, x_2\neq 0\}$;

(d) $x_1^2e^{x_2}-x_1\sin x_2$, $D=\mathsf{R}^2$.

18. $\dfrac{du}{dx_1}=\dfrac{-(u^2-1)}{x_1(3-u)}$; $(x_1-x_2)=C(x_1+x_2)^2$. Introduce new coordinates $y_1=x_1+x_2$ and $y_2=x_1-x_2$ and plot curves for $C=0, \pm 1, \pm 2, \pm 3$.

19. The trajectories of both systems lie on the solutions of the differential equation $\dfrac{dx_2}{dx_1}=-\dfrac{x_2}{x_1}$. The first system has a line of fixed points ($x_1=x_2^2$) whereas the second system has a saddle point at the origin.

20. $x_2^2=(\ln x_1)^2+C$.

21. $J=\begin{bmatrix}\lambda_1 & 0\\ 0 & \lambda_2\end{bmatrix}$, $(x_2/x_1)^{\lambda_2/\lambda_1}$, $D=\mathsf{R}^2\backslash\{x_2\text{-axis}\}$

$J=\begin{bmatrix}0 & -\beta\\ \beta & 0\end{bmatrix}$, $x_1^2+x_2^2$, $D=\mathsf{R}^2$.

$J=\begin{bmatrix}\alpha & -\beta\\ \beta & \alpha\end{bmatrix}$, $re^{-\alpha\theta/\beta}$, $\beta\neq 0$ $D=\mathsf{R}^2\backslash\{0\}$.

Let the trajectory $\mathbf{x}(t)$ tend to a fixed point $\mathbf{x}_0$ as $t\to\infty$. A first integral f satisfies $f(\mathbf{x}(t))=f(\mathbf{x}_0)$, by continuity of f. If $\mathbf{x}_0$ is asymptotically stable then $f(\mathbf{x})=f(\mathbf{x}_0)$ *for all* $\mathbf{x}$ in some neighbourhood of $\mathbf{x}_0$.

22. $H=\frac{1}{2}x_2^2+\dfrac{x_1^2}{2}-\dfrac{\alpha x_1^3}{3}$, centre at (0, 0), saddle at $(\alpha^{-1}, 0)$.

24. (a) Let $x_2(0)\geqslant 0$; then $\dot{x}_2\geqslant 0$ implies $x_2(t)\geqslant x_2(0)$ for all positive t;

(b) For $x_2=\beta x_1$, $\dot{x}_1=(\beta-\alpha)x_1$ and $\dot{x}_2=(\beta-\alpha)x_2$. Thus $\dot{x}_2=\beta\dot{x}_1$;

(c) In polar coordinates $\dot{r}=-r(1-r^2)$ and hence $\dot{r}<0$ for $r<1$;

(d) The trajectories lie on the family of hyperbolae $x_1x_2=C$. Show that the boundary of the indicated region is a parabola of fixed points.

25. (a) $2\dot{x}^2 + x^4 = 1$; (b) $\dot{x}^2 + x^2 = 5$; (c) $2\dot{x}^2 + x^4 = 2$.

26. Yes.

27. If r is the radial distance, $\dot{r} = r(1 - r^2) - F \sin\theta$. When $F = 0$, the limit cycle is given by $r \equiv 1$.

28. The system has no fixed points. If there exists at least one limit cycle choose a region R bounded by such a trajectory which contains no other limit cycles. Use Poincare–Bendixson theory to obtain a contradiction by showing a fixed point should exist in R.

29. $\dfrac{\partial X_1}{\partial x_1} + \dfrac{\partial X_2}{\partial x_2} = 2 - 4(x_1^2 + x_2^2)$. A limit cycle either intersects $x_1^2 + x_2^2 = \frac{1}{2}$ or contains this curve in its bounded region.

30. Locate the subregions of R of constant sign in $\dot{x}_1$ and $\dot{x}_2$. Consider the behaviour of the trajectories through the points on the parabola $x_2 = -x_1^2 + 3x_1 + 1$, and show that they spiral in an anticlockwise sense about the fixed point.

Chapter 4

3. The principal directions are $(1, \lambda_1)$, $(1, \lambda_2)$ where λ_1, λ_2 $(\lambda_1 > \lambda_2)$ are the eigenvalues. Show that as $k \to \infty$, $\lambda_1 \to 0$ and $\lambda_2 \to -\infty$.

4. $\dfrac{d^2 i}{dt^2} + \dfrac{1}{CR}\dfrac{di}{dt} + \dfrac{1}{CL} i = 0.$

5. $L\dfrac{dj_R}{dt} + Rj_R = E_0.$

6. $C\dfrac{dv_C}{dt} = \dfrac{v_C}{R}.$

8. $\dfrac{D}{U+D} = \dfrac{kD_0}{(k+\mu)D_0 + (kU_0 - \mu D_0)e^{-kt}}, \; k = \lambda - \nu - \mu.$

$$\frac{D}{U+D} \to \frac{k}{k+\mu} \text{ as } t \to \infty.$$

9. $x = \dfrac{1}{\omega}\sin \omega t, \; y = \cos t.$

10. $\boldsymbol{M} = \begin{bmatrix} 1 & 1 \\ 1 & -1 \end{bmatrix}, \quad \boldsymbol{D} = \begin{bmatrix} -k/m & 0 \\ 0 & -3k/m \end{bmatrix}.$

Normal modes: $y_1 \equiv 0$, the particles undergo oscillations that are symmetric relative to the mid-point of AB; $y_2 \equiv 0$, oscillations occur with particles fixed at a distance l apart.

11. (a) The equilibrium state is $\left(\frac{\alpha G_0}{\alpha-1}, \frac{G_0}{\alpha-1}\right)$; introduce local coordinates and use Trace-Det classification.

(b) The equilibrium state is $\left(\frac{\alpha G_0}{[\alpha(1-k)-1]}, \frac{G_0}{[\alpha(1-k)-1]}\right)$ with $k < \alpha - 1/\alpha\ (= A)$. The stable equilibrium point moves off to infinity as $k \to A$.

13. $A = \begin{bmatrix} 0 & 1 \\ -1 & -5/2 \end{bmatrix}$, $e^{A\tau} = \frac{1}{3}\begin{bmatrix} 4e^{-\frac{1}{2}\tau} - e^{-2\tau} & 2e^{-\frac{1}{2}\tau} - 2e^{-2\tau} \\ -2e^{-\frac{1}{2}\tau} + 2e^{-2\tau} & -e^{-\frac{1}{2}\tau} + 4e^{-2\tau} \end{bmatrix}$.

14. The steady state solutions are:

$$i_1 = \frac{E_0}{\sqrt{(R^2+\omega^2 L^2)}}\cos(\omega t - \phi), \text{ where } \tan\phi = \omega L/R;$$

$i_2 = -E_0 C\omega \sin\omega t.$

The current $i_1 + i_2$ has amplitude

$$\frac{E_0 C}{L}\left[R^2 + \left(\omega L - \frac{1}{\omega C}\right)^2\right]^{\frac{1}{2}}.$$

15. The fixed points are at (a) (0, 0), unstable node; (b) (2, 0), stable improper node; (c) (0, 2), stable improper node; (d) (2/3, 2/3), saddle. The principal directions are (a) (1, 0) and (0, 1); (b) (1, 0); (c) (0, 1); (d) (1, 1) and (1, −1).

16. The fixed points occur at $0 = (0, 0)$, $A = (0, v)$, $B = (1, 0)$ and $C = ((1-v)/(1-4v^2),\ v(1-4v)/(1-4v^2))$ and depend on v as follows:

$$A: \begin{cases} \text{saddle } 0 < v < 1 \\ \text{stable node } v > 1 \end{cases}$$

$$B: \begin{cases} \text{saddle } v < \frac{1}{4} \\ \text{stable node } v > \frac{1}{4} \end{cases};$$

$$C: \begin{cases} \text{stable node } v < \frac{1}{4} \\ \text{saddle } v > 1 \end{cases}$$

(C is not in the first quadrant for $\frac{1}{4} < v < 1$)

O: unstable node $v > 0$.

17. Fixed points at $0 = (0, 0)$, $A = (0, -1/\alpha)$, $B = (1/\alpha, 0)$, $C = \left(\frac{1+\alpha}{1+\alpha^2}, \frac{1-\alpha}{1+\alpha^2}\right)$. Linearization at C has trace $-2\alpha/(1+\alpha^2)$, determinant $(1-\alpha^2)/(1+\alpha^2)$ and hence $\Delta < 0$.

18. $\bar{x}_1 = \frac{1}{T}\int_0^T x_1 \mathrm{d}t = \frac{1}{T}\int_0^T \left(\frac{c}{d} + \frac{\dot{x}_2}{x_2}\right)\mathrm{d}t = \frac{c}{d}.$

$\int_{x_2(0)}^{x_2(T)} \frac{\mathrm{d}x_2}{x_2} = 0$, since $x_2(0) = x_2(T)$.

$\bar{x}_2 = a/b$.
With harvesting $\dot{x}_1 = x_1[(a-\varepsilon) - bx_2]$,
$\dot{x}_2 = -x_2[(c+\varepsilon) - dx_1]$,
$\bar{x}_1 = \frac{c+\varepsilon}{d}, \bar{x}_2 = \frac{a-\varepsilon}{b}.$

19. The peak of the parabola occurs at $y_1 = \frac{k-d}{2}$ which is greater than 1. Use this to deduce that $\mathrm{Tr}(\boldsymbol{W})$ (see (4.94)) is negative. A phase portrait with just one (stable) limit cycle must contain an unstable fixed point.

20. $\dot{B} = \gamma \dot{P}$ when $B = \gamma P$.
For $\mu(P) = b + cP$, fixed points are: $0 = (0, 0)$, unstable node; $S = (-b/c,\ 0)$, saddle; $T = \left(\frac{\gamma - b}{c}, \frac{\gamma(\gamma - b)}{c}\right)$, stable node or focus. The fixed point T has the positive quadrant as a domain of stability.

21. $y = -x + \frac{1}{2}\ln x - c_0$ (orientation is given by x decreasing for $x, y > 0$). The number of susceptibles (x) decreases and the infectives (y) increase to a maximum before decreasing to zero.

22. The fixed point $(\frac{1}{2}, 1)$ is a stable focus. The epidemic sustains a non-zero number of infectives.

23. $I = 1 - S + \frac{1}{\sigma}\log(S/S_0)$ (at $t = 0$, $I_0 + S_0 = 1$, $R_0 = 0$).

(a) $\sigma S_0 \leqslant 1$, then $\sigma S(t) < 1$ for all positive t ($\dot{S} < 0$).
Hence $\dot{I} = \gamma I(\sigma S - 1)$ is negative and $\dot{I} = 0$ if and only if $I = 0$.

(b) $\sigma S_0 > 1$, S decreases and so let $t = t_0$ satisfy $\sigma S(t_0) = 1$.
Then $\dot{I}$ is positive for $t < t_0$ and negative for $t > t_0$.

Note $S = S_L$ when $I = 0$. To show that S_L is unique, prove that I is an increasing function of S in $(0, 1/\sigma)$ where $I(0) < 0$ and $I(1/\sigma) > 0$.

24. A first integral is $f(X_1, Y_2) = h(X) + g(Y)$ where $h(X) = \frac{1}{2}\alpha X^2$

$-\beta X$ and $g(Y) = bY - \dfrac{a}{k}\ln(A + kY)$. The functions $h(X)$ and $g(Y)$ both have global minima for positive values of X and Y respectively.

25. A first integral is $f(x_1, x_2) = \dfrac{x_2^2}{2} + \omega_0^2(1 - \cos x_1)$. The linearized system at the fixed points $(2n\pi, 0)$, n integer, is a centre and at the fixed points $((2n+1)\pi, 0)$, n integer, is a saddle.

26. At the fixed point $(0, \nu^{-1})$, eigenvalues are $-\mu + \dfrac{1}{\nu}$, $-\nu$; at the fixed point $(\pm\sqrt{(1-\mu\nu)}, \mu)$, the trace is $-\nu$ and the determinant is $2(1-\mu\nu)$.

27. $r\dot{r} = x_1\dot{x}_1 + x_2\dot{x}_2$; $r^2\dot{\theta} = x_1\dot{x}_2 - x_2\dot{x}_1$; $\dot{r} = \varepsilon r_0 \sin^2\theta(1 - r_0^2 \cos^2\theta) + 0(\varepsilon^2)$; $\dot{\theta} = -1 + \varepsilon(r_0^2\cos^2\theta - 1) + 0(\varepsilon^2)$. $\Delta r = \varepsilon\pi r_0(1 - r_0^2/4)$.
$\Delta r > 0$, $r_0 < 2$, $\Delta r < 0$, $r_0 > 2$, to first order in ε and so $\Delta r = 0$ for $r_0 = 2$.

28. (a) Obtain the second order equations in x and y given by the system $\dot{x} = y - \varepsilon(\frac{1}{3}x^3 - x)$, $\dot{y} = -x$.

(b) If $x_2 = \varepsilon\omega$, then $\dot{x}_1 = \varepsilon(\omega - \frac{1}{3}x_1^3 + x_1)$, $\dot{\omega} = -\dfrac{X_1}{\varepsilon}$. As $\varepsilon \to \infty$, $\dot{x}_1 \to \infty$ except near $\omega = \frac{1}{3}x_1^3 - x_1$.

29. Introduce a capacitor C in parallel; the circuit equations become $C\dfrac{dv_C}{dt} = j_L - v_C^3 + v_C$, $L\dfrac{dj}{dt} = -v_C$. The system equations are of Liénard type and hence the circuit oscillates.

30. $T_1 = \displaystyle\int_q^Q \frac{dp}{(ap - bp^2)}$, $T_2 = \displaystyle\int_Q^q \frac{dp}{(Ap - Bp^2)}$.

The population cycle has period $T_1 + T_2$.

31. Obtain the system equations
$\ddot{Y} + k\dot{Y} + Y = L \quad (\dot{Y} > 0)$,
$\ddot{Y} + k\dot{Y} + Y = -M \quad (\dot{Y} < 0)$.
Use the substitution $Y' = Y - L$ in both equations to draw an analogy with the system considered in Section 4.5.2.

32. $Y_1(t) = G_0 t e^{1-t}$, $0 \leqslant t < 1$, $Y_1(t) = G_0$, $t \geqslant 1$. The general solution of the differential equation for $t > 1$ is $Y(t) = (At + B)e^{-t} + G_0$, where A and B are constants. Regardless of the choice of A and B, $Y(t) \to G_0$ as $t \to \infty$.

Chapter 5

1. (d) $\dot{V}(x_1, x_2) = -2x_1^2(\sin x_1)^2 - 2x_2^2 - 2x_2^6$ is negative definite when $x_1^2 + x_2^2 < \pi^2$.

(e) $\dot{V}(x_1, x_2) = -2x_1^2(1 - x_2) - 2x_2^2(1 - x_1)$ is negative definite when $x_1^2 + x_2^2 < 1$.

2. The domain of stability is R^2 for (a), (b) and (c) and $\{(x_1, x_2) \mid x_1^2 + x_2^2 < r^2\}$ where $r = \pi$ for (d) and $r = 1$ for (e).

3. Asymptotically stable: (a) and (b).
Neutrally stable: (c) and (d).

4. The system $\dot{\mathbf{x}} = -\mathbf{X}(\mathbf{x})$ has an asymptotically stable fixed point at the origin. Let $\mathbf{x}_0$ be such that $\lim_{t \to \infty} \phi_t(\mathbf{x}_0) = \mathbf{0}$. Choose a neighbourhood N of $\mathbf{0}$ not containing $\mathbf{x}_0$. The trajectory through $\mathbf{x}_0$ of the system $\dot{\mathbf{x}} = \mathbf{X}(\mathbf{x})$ satisfies $\lim_{t \to -\infty} \phi_t(\mathbf{x}_0) = \mathbf{0}$. Use this property to show that the origin is unstable. Use the function $V(x_1, x_2) = x_1^2 + x_2^2$ in (a) to (c).

5. Use the function $V(x_1, x_2) = x_1^2 + x_2^2$ in (a) and (b) and $V(x_1, x_2) = x_1^4 + 2x_2^2$ in (c) and (d).

6. If V is positive definite then $V(1, 0)$ is positive and so a is positive; also

$$V(x_1, x_2) = a\left(x_1 + \frac{b}{a}x_2\right)^2 + \left(c - \frac{b^2}{a}\right)x_2^2$$

and thus a and $c - b^2/a$ are positive. Try $a = 5, b = 1, c = 2$; then

$$V(x_1, x_2) = 5\left(x_1 + \frac{x_2}{5}\right)^2 + \frac{9}{5}x_2^2.$$

For $V(x_1, x_2) < 9/5, x_2^2 < 1$ and so there is a domain of stability defined by $25x_1^2 + 10x_1x_2 + 10x_2^2 < 9$.

7. (a) $V(x_1, x_2) = x_1^2 + x_2^2$, $\dot{V}(x_1, x_2) = -2r^2(1 - r^2)(1 + r^2)$; $x_1^2 + x_2^2 < 1$.

(b) $V(x_1, x_2) = x_1^6 + 3x_2^2$, $\dot{V} = -x_2^2(1 - x_2^2)$; $x_1^6 + 3x_2^2 < 3$.

8. $\dot{V}(x_1, x_2) = \dfrac{2x_1^2}{a^2}(x_1 - a) + \dfrac{2x_2^2}{b^2}(x_2 - b)$ is negative definite for $\dfrac{x_1^2}{a^2} + \dfrac{x_2^2}{b^2} < 1$.

9. $V(x_1, x_2) = \frac{1}{2}x_1^4 + \frac{1}{2}x_1^2 - x_1x_2 + x_2^2$ satisfies the hypotheses of Theorem 5.4.3.

10. $\dot{V}(x_1, x_2) = 3(x_1^2 + x_2^2)^2$.

11. $V(x_1, x_2) = x_1 - x_2$.
One separatrix is $x_2 = 0$ where $\dot{x}_1 = x_1^4$. Thus the phase portrait on the x_1-axis is a shunt and so the fixed point is unstable.

12. (a) $\mu < 0$: unstable focus, stable limit cycle at $r = -\mu$;
$\mu \geqslant 0$: stable focus.
(b) $\mu < 0$: stable focus, semi-stable limit cycle at $r = -\mu$;
$\mu = 0$: centre;
$\mu > 0$: unstable focus,
(c) $\mu \leqslant 0$: unstable focus;
$\mu > 0$: unstable focus, stable limit cycle at $r = \mu/2$ and unstable limit cycle at $r = \mu$.
(d) $\mu \leqslant 0$: stable focus;
$\mu > 0$: unstable focus surrounded by a stable limit cycle at $r = \sqrt{\mu}$.
(e) $\mu < 0$: stable focus (clockwise);
$\mu = 0$: plane of fixed points;
$\mu > 0$: unstable focus (anti-clockwise).
(f) $|\mu| < 1$: unstable focus (anti-clockwise);
$|\mu| = 1$: unstable star node;
$|\mu| > 1$: unstable focus (clockwise).

13. (a) $I = -4$; (b) $I = -2$; (c) $I = -2$.

15. Let $\dot{x}_1 = x_2$, $\dot{x}_2 = -x_1 + \mu x_2 - x_2^3$. The system bifurcates from a stable fixed point to stable limit cycles surrounding an unstable fixed point as μ passes through 0.

16. Use $\boldsymbol{x} = \begin{bmatrix} 1 & -2 \\ 1 & -1 \end{bmatrix} \boldsymbol{y}$ to obtain $\dot{y}_1 = \mu y_1 + y_2$, $\dot{y}_2 = -y_1 + \mu y_2 - y_2^3$.

17. Consider the system which has trajectories with reverse orientation. Show that this system undergoes a Hopf bifurcation to *stable* limit cycles at $\mu = 0$.

Index